Artificial Intelligence Oversight and Cybersecurity

Protecting Systems, People, and Trust Through Governance, SOC Transformation, and Resilient Security Architecture

Ozwald Carter

Cybersoft Publishing LLC

Fort Washington, MD 20744

Drclaude.net

Second Edition March 2026

Table of Contents

Introduction

The threat landscape you managed five years ago no longer exists. What replaced it is something far more consequential: an adversary environment supercharged by artificial intelligence, where attacks launch faster than human analysts can respond, where deception operates at machine scale, and where the cost of mounting a sophisticated campaign has collapsed to near zero. Senior leaders who once delegated cybersecurity entirely to technical teams discover that AI has dissolved the boundary between a technology problem and a leadership problem.

AI-powered attacks have rewritten the economics of the offense. Automated reconnaissance tools now map an organization's full digital surface in hours, identifying misconfigured systems, exposed credentials, and third-party dependencies with a thoroughness no human attacker team could match. Spear-phishing campaigns that once demanded skilled social engineers are now generated at scale by large language models that craft individualized, contextually credible lures. The barrier to entry for a devastating intrusion has never been lower, and the volume of attempts has never been higher.

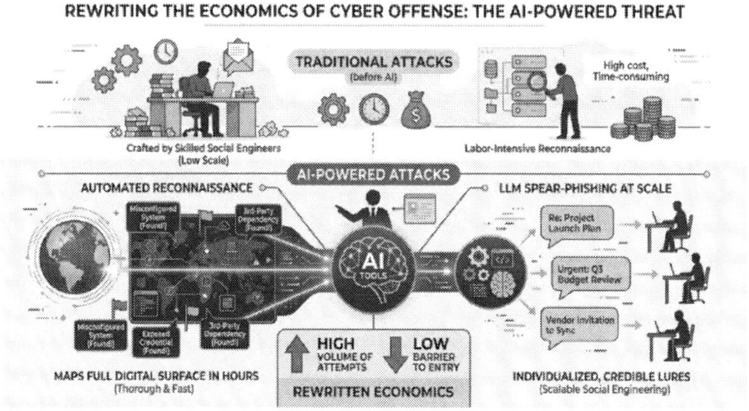

Deepfakes and synthetic media have introduced a dimension of threat that strikes at something more fundamental than data: organizational trust itself. Executives' voices are cloned to authorize fraudulent wire transfers. Video fabrications circulate to discredit institutions or manipulate markets. Intelligence agencies and criminal organizations alike now use synthetic identity attacks to infiltrate supply chains and bypass identity verification systems that were once considered robust. When the authenticity of communications can no longer be assumed, the entire fabric of organizational decision-making is at risk.

Adaptive malware represents perhaps the most technically alarming frontier. Traditional signature-based defenses assume an attacker's tools remain static long enough to be identified and blocked. AI-enabled malware mutates its behavior mid-campaign, learns from detection attempts, and routes around defenses in real time. Ransomware strains now include reinforcement-learning components that study an organization's backup architecture before triggering encryption, timing their strike to maximize damage and ransom leverage. The attacker is no longer just faster than the defender; in many environments, the attacker is actively optimizing against the defender's specific posture.

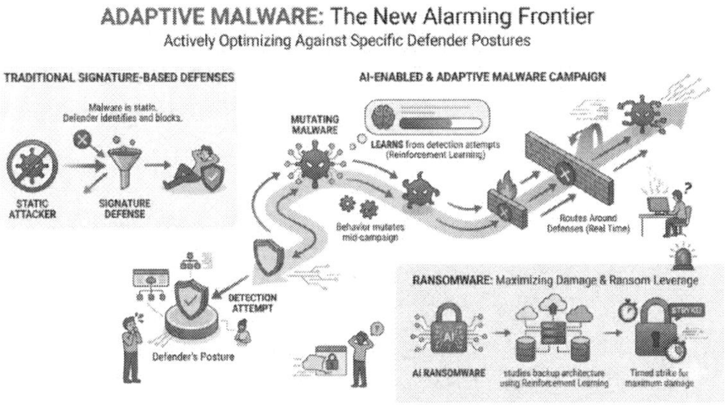

This book is written for leaders who understand that these dynamics cannot be managed from a distance. The defender-attacker asymmetry created by AI demands that senior managers engage directly with questions of strategy, investment, governance, and organizational design. The chapters ahead offer both the conceptual grounding and the practical frameworks needed to lead an enterprise through this era with clarity and confidence. The goal is not to turn executives into security engineers, but to ensure that the people responsible for enterprise resilience are equipped to ask the right questions, make the right bets, and build the right teams.

This book is organized as a guided progression through the AI cybersecurity landscape, moving from understanding the threat to building the organizational capacity to meet it. Each chapter is designed to stand on its own while contributing to a cumulative picture of what it means to defend an enterprise in an era defined by intelligent, adaptive adversaries. Senior managers will find that each topic connects directly to decisions they are already being asked to make about budget, workforce, vendor selection, and risk tolerance.

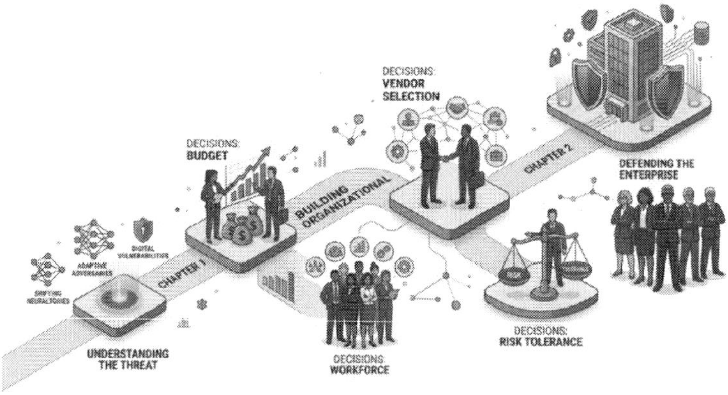

The journey begins with the adversary. Chapters One and Two examine AI from both sides of the conflict: how attackers have deployed it to accelerate and scale their operations, and how defenders are deploying it to detect, classify, and respond to threats that would overwhelm human analysts working alone. Understanding both sides is essential because the most effective defensive investments are those calibrated to the adversary's specific AI capabilities.

The middle section of the book confronts attack surfaces and deception. Chapter Three maps the vulnerabilities that AI systems themselves introduce into enterprise environments. Chapter Four examines the problem of trust destruction posed by deepfakes and synthetic media, with particular attention to the organizational and governance implications for leaders who must decide how their institutions communicate and verify identity. Chapter Five then turns the lens inward, addressing how to secure AI systems against manipulation, poisoning, and exfiltration.

Chapters Six, Seven, and Eight address the organizational transformation that effective AI-powered defense requires. Chapter Six examines the evolving security operations center and the role AI plays in managing alert volume and analyst capacity. Chapter Seven addresses governance: the policy frameworks, regulatory expectations, and ethical boundaries that must guide how organizations deploy AI in defensive roles. Chapter Eight focuses on the people dimension, exploring how security teams must evolve, what skills will define the next generation of security professionals, and how human-AI teaming models are changing the nature of the work.

The final chapter and closing synthesize the preceding material into a forward-looking architecture for enterprise resilience. Readers will find there a framework for continuous adaptation, a model for integrating threat intelligence into strategic planning, and a leadership posture that treats cybersecurity not as a compliance obligation but as a core competency. Throughout every chapter, the emphasis is on decisions senior managers can actually make, not on technical depth for its own sake.

1 The Machine That Learned to Attack

At 3:47 on a Tuesday morning, a network monitoring alert fired inside a mid-sized federal agency outside Washington, D.C. The on-call analyst assumed it was another false positive from the agency's aging intrusion-detection stack. By the time the day shift arrived, attackers had already mapped 17 internal systems, identified 4 lateral movement paths, extracted credential hashes from 2 domain controllers, and staged ransomware payloads in 3 dormant service accounts. The entire operation took less than six hours. Post-incident forensics revealed that the adversary's toolkit included an AI-assisted reconnaissance engine purchased from a dark web marketplace for less than $3,000. The tool performed, in six hours, what a skilled human penetration tester typically requires several days to accomplish.

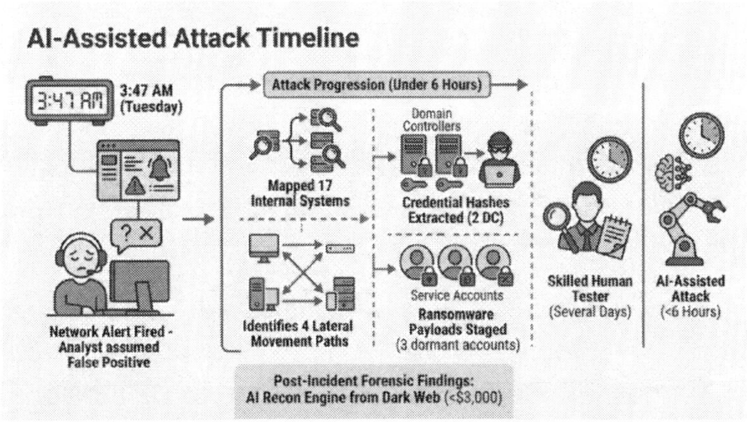

This is not a hypothetical scenario. Variants of it are appearing in incident reports from federal contractors, critical infrastructure operators, and large financial institutions with increasing regularity. The threat landscape has not simply grown larger; it has fundamentally changed its character. Attackers now use tools that learn, adapt, and optimize to counter the specific defenses they encounter. The asymmetry that has always favored attackers, a defender must protect everything. In contrast, an attacker needs only to find one gap, which machine intelligence has sharpened into something qualitatively different from anything security organizations have faced before.

1.1 Why It Matters

Understanding how AI has changed the offensive side of the threat equation is not an academic exercise. It is a prerequisite for rational security investment. Every dollar a security manager allocates to defensive controls, every policy reviewed by general counsel, and every briefing prepared for the board is premised on a model of how attackers operate. If that model is outdated, if it assumes human-speed reconnaissance, manually crafted phishing emails, and static malware, then the defenses derived from it will be systematically miscalibrated. Budget will flow to the wrong places, and the board will be asking the wrong questions.

The practical stakes are immediate. AI-assisted attacks compress the window between initial compromise and material damage. Traditional detection strategies that rely on dwell time catching attackers during the days or weeks they spend moving laterally before executing their final objective are undermined when dwell time shrinks from weeks to hours. The incident-response playbook that assumed a sixty-hour detection window becomes nearly useless when the adversary has already achieved their objective before the first analyst opens a ticket.

For security managers, this creates a direct workflow implication: risk assessments must be re-run against a faster adversary model, detection investments must prioritize speed over comprehensiveness, and communication with senior leadership must convey a threat environment whose pace has changed, not just its sophistication. None of those adjustments can be made without a clear-eyed understanding of what AI-augmented offensive tools actually do.

Diagram 1.1: AI-Augmented Attack Timeline

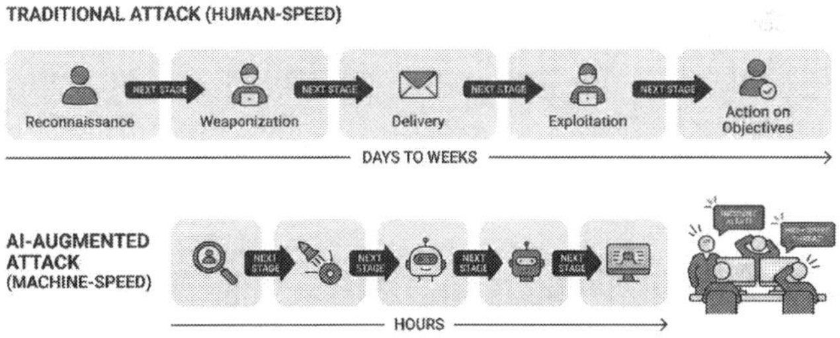

AI enables attackers to accelerate the cyberattack lifecycle, completing complex stages in hours rather than days.

1.2 Automated Reconnaissance and Attack Planning

How AI Tools Map an Organization's Entire Attack Surface in Hours

Reconnaissance has always been the first phase of any serious attack: gather information, identify targets, and map the terrain before moving. Traditionally, it required patient human labor to scan ports, analyze certificates, correlate job postings with technology stacks, and trace organizational structures on LinkedIn. Each of those tasks is individually tractable but collectively time-consuming. Human attackers would spend days assembling a target profile that was still incomplete by the time they moved to exploitation.

AI-assisted reconnaissance tools change this in two ways. First, they automate the correlation of data sources that a human analyst would address sequentially. A modern offensive reconnaissance engine can simultaneously query DNS records, certificate transparency logs, passive DNS databases, code repositories, paste sites, and social media, then run automated inference across all of those feeds to assemble a target profile. What took a skilled attacker two days can be reduced to two hours. Second, and more importantly, machine learning components can prioritize attack-surface elements based on estimated exploitability. Rather than delivering a raw data dump, these tools score targets exposed services, unpatched systems, misconfigured cloud buckets, accounts with weak credential signals, and surface the highest-value entry points first.

The implication for defenders is that the assumption of time to discover as a protective buffer has eroded significantly. A public-facing misconfiguration that would previously have gone unnoticed for months, because no attacker had gotten around to scanning it, can now be identified and queued for exploitation within hours of introduction. Attack surface management must operate at the same speed as automated offensive scanning, which means continuous, automated, and integrated with vulnerability prioritization, not periodic manual reviews.

The speed advantage of AI-assisted reconnaissance also compounds over multiple campaigns. When an adversary deploys automated tools against hundreds of targets simultaneously, they accumulate environment-specific intelligence that refines their attack approach for each target. An organization that was scanned six months ago has already contributed to the attacker's understanding of its industry's common infrastructure patterns, typical exposed services, and likely patch cadence. Security managers should assume that any organization of meaningful size has already been profiled by automated reconnaissance tools, regardless of whether a formal attack has been attempted.

Supply Chain and Third-Party Dependency Exploitation at Scale

Sophisticated attackers have long understood that the most valuable targets are often best reached through their suppliers, partners, and technology vendors rather than directly. The 2020 SolarWinds breach demonstrated at scale what targeted supply chain attacks could achieve. AI changes this threat in a specific way: it makes supply chain reconnaissance tractable at scale. Mapping the dependency graph of a target organization, which software vendors feed into their environment, which cloud services process their data, and which managed service providers have privileged network access, is now something an automated system can assemble from public sources in a fraction of the time it would take a human analyst.

For security managers, this means that the organization's effective attack surface includes not just its own systems but the systems of every entity that has privileged access to its environment. AI tools enable adversaries to find the weakest link in that extended chain automatically. A vendor with a legacy VPN concentrator, a partner with inadequate MFA enforcement, and a SaaS platform with a disclosed but unpatched API vulnerability, each becomes a potential entry point that an automated reconnaissance engine will identify and score before human attackers ever engage.

From Open-Source Intelligence to Precision Targeting

Open-source intelligence OSINT has always been part of the attacker's toolkit, but it required skill, patience, and experience to turn raw information into actionable targeting. AI dramatically lowers that skill floor. Modern LLM-augmented OSINT tools can process thousands of documents earnings calls, regulatory filings, press releases, LinkedIn profiles, conference talks and synthesize from them a detailed organizational map: who holds budget authority, which systems are being deprecated or migrated, where the security team is stretched thin, which executives travel frequently and therefore face physical access risks at hotel networks and airport lounges.

Precision targeting converts broad reconnaissance into personalized attacks. When an adversary knows that a specific finance director is managing a system migration, that they attended a particular conference last month, that they report to a CFO whose name and email format are publicly known, and that their organization uses a specific ERP platform with a known vulnerability, they have everything needed to craft an attack that is tailored, timely, and highly likely to succeed. No element of that profile required sophisticated hacking to obtain; it was all assembled from public sources by a machine learning system running overnight.

Diagram 1.2: OSINT-to-Precision-Target Pipeline

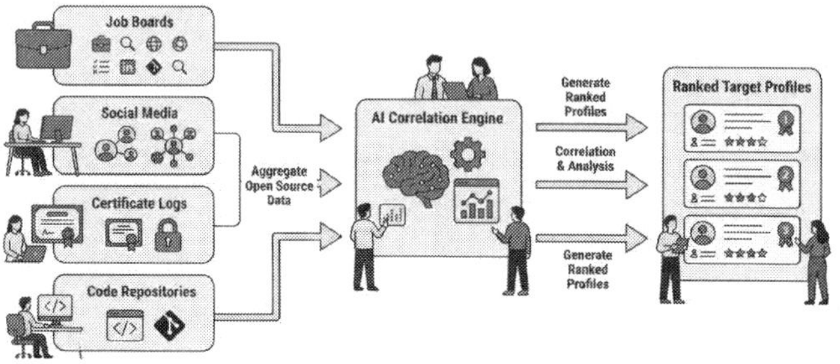

This pipeline automates OSINT collection and AI analysis to generate prioritized target profiles for cybersecurity operations.

Automated OSINT aggregation also extends to technical signals that human analysts would rarely correlate manually. Certificate issuance patterns can reveal system staging timelines; GitHub commit histories can expose internal infrastructure naming conventions and configuration management approaches; Shodan and similar services maintain persistent inventories of internet-exposed services that update faster than most organizations' own asset management systems. AI tools that integrate these signals produce a more comprehensive target picture than most organizations maintain of their own attack surface, a fundamentally asymmetric situation that continuous attack surface management programs are designed to correct.

1.3 AI-Generated Phishing and Social Engineering

Large Language Models as Spear-Phishing Engines

Phishing has been the most reliably effective initial access vector in enterprise breaches for more than a decade. The reason is straightforward: it targets humans, who are consistently harder to patch than software. AI has not invented phishing; it has made phishing dramatically more scalable and more convincing, simultaneously a combination that was previously impossible. Historically, sophisticated spear-phishing required significant human effort: researching the target, crafting a personalized message, and timing the delivery appropriately. That labor constraint limited volume. Automated phishing at scale produced generic lures that trained users could identify. AI breaks this tradeoff.

A large language model can ingest a target profile assembled by the reconnaissance tools described above and generate a personalized, contextually accurate phishing email indistinguishable in tone and content from legitimate correspondence. It can reference the specific project the recipient is working on, mirror the communication style of a known colleague, include plausible urgency tied to a real organizational event, and calibrate the level of technical detail to match the recipient's role. The same system can do this for ten thousand targets simultaneously. Volume and quality, which were previously in tension, are now both achievable simultaneously.

For security managers, awareness training faces a harder challenge than it did three years ago. The heuristics users have been taught look for generic greetings, poor grammar, mismatched sender domains, and implausible urgency were calibrated against the phishing attacks of the prior era. An AI-generated spear-phishing message will have a correct greeting, flawless grammar, a spoofed sender domain with a credible lookalike, and urgency tied to something the recipient actually recognizes as real. User training must be updated, but more importantly, technical controls that do not depend on user judgment must be prioritized.

Contextual Personalization: Beyond Generic Lures

The distinguishing feature of AI-assisted social engineering is not just grammatical correctness but contextual coherence. An attacker who knows that a target company is currently migrating to a new HR platform can generate phishing lures that reference that migration, instructing recipients to re-authenticate their credentials on a spoofed portal. The message will contain accurate details about the HR system, reference the migration project timeline as publicly signaled in job postings, and use language consistent with how the HR department communicates. The recipient has no reason to be suspicious because every element of the message matches reality.

The economic calculus of AI-assisted phishing has changed the threat landscape in ways that affect organizations that have previously relied on their relative obscurity or modest profile to limit targeting. When spear-phishing required hours of manual research per target, adversaries were selective, targeting only organizations whose access warranted the investment. When AI tools can process a thousand targets per hour, selectivity is no longer a constraint. Mid-sized organizations, regional government agencies, educational institutions, and healthcare providers that once benefited from being below the threshold of attacker effort are now within the operating range of automated campaigns. The threat is no longer reserved for large corporations and federal agencies.

This contextual coherence extends to personalization of voice and communication channels. AI tools can generate not just email text but SMS messages, voicemail scripts, and even real-time voice synthesis that mimics a known colleague or manager. Multi-channel attack sequences, a phishing email followed by a follow-up SMS from a spoofed number, validated by a synthetic voice call, exploit the human tendency to trust information that arrives through multiple independent channels. Each channel appears to confirm the others, creating a reinforcing illusion of legitimacy.

Multi-Channel Attack Sequences and Conversation Hijacking

Some of the most sophisticated AI-assisted social engineering attacks do not rely on initial deception but on hijacking ongoing legitimate conversations. An attacker who has compromised a supplier's email account, or who has assembled enough OSINT to reconstruct the content of ongoing business conversations, can use an AI system to generate reply-chain messages that appear to be seamless continuations of legitimate threads. The recipient does not receive a cold phishing email; they receive what appears to be a follow-up to a conversation they were already having.

Conversation hijacking attacks are particularly dangerous because they undermine the heuristic most users use to verify legitimacy: checking whether the request is consistent with an established relationship. When the attacker has reconstructed or hijacked an established relationship context, that check returns a false positive. Technical controls that rely on behavioral signals such as first-time sender detection, lookalike domain flagging, and attachment sandboxing may also fail when the attack originates from a legitimately compromised account within a trusted domain.

Diagram 1.3: Multi-Channel AI Phishing Sequence

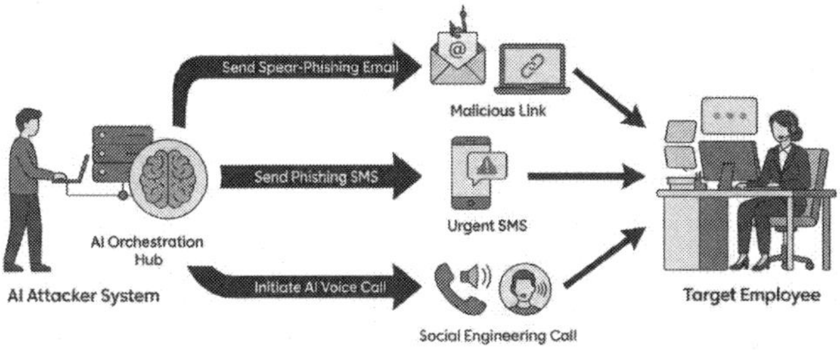

An advanced AI attacker concurrently employs email, SMS, and voice calls to increase the probability of compromising a target employee.

1.4 Adaptive and Self-Modifying Malware

Reinforcement Learning in Offensive Toolkits

Classical malware was static: once compiled and deployed, it behaved the same way regardless of what it encountered. Defenders could characterize their behavior, build signatures, and share those signatures across organizations. The signature-based detection paradigm that dominated endpoint security for two decades was premised on this static character. AI-augmented malware challenges the premise at its foundation.

Reinforcement learning frameworks enable offensive tools to treat the target environment as a problem to optimize against. The malware takes an action that attempts a particular privilege escalation technique, tries to disable a specific monitoring agent, tests a lateral movement path, and observes whether that action produces a reward signal (success, partial success, or failure), and updates its behavior policy accordingly. Over multiple interactions, the tool converges on a sequence of actions that achieves the attacker's objective while minimizing the probability of triggering defensive responses. Each deployment becomes more effective than the last as the tool accumulates experience across environments.

This creates a moving target problem for defenders. A detection rule that successfully identifies a specific behavior pattern on Monday may fail on Friday because the adversarial system has learned to avoid that pattern. The traditional cycle of incident, forensics, signature, and deployment, which can be completed in hours for static malware, becomes inadequate when the malware itself evolves in response to the defensive environment it encounters.

Evasion Optimization Against Specific Defensive Postures

Beyond general reinforcement learning, specialized evasion-optimization tools have emerged that target specific defensive products. By probing a target environment with benign traffic and observing responses, which actions trigger alerts, which are logged but not alerted, which pass completely undetected, an AI tool can build a model of the defensive posture and generate attack traffic optimized to fall below detection thresholds. This is essentially adversarial machine learning applied to the problem of penetrating a specific organization's defensive stack.

The practical implication for security managers is that standard benchmarking of defensive tools against known attack samples is no longer sufficient. A detection model that performs admirably on the MITRE ATT&CK evaluation dataset may perform poorly against an adversary that has spent time probing its detection boundaries and crafting traffic specifically designed to evade it. Red team exercises must include AI-augmented adversary simulation, and vendor evaluations must include adversarial robustness testing rather than recall against known samples.

Timing and Environmental Awareness in AI-Driven Ransomware

AI-augmented ransomware represents a particularly consequential application of these capabilities. Sophisticated ransomware operators have long understood that the timing of encryption matters: execute too early, and you are discovered before maximum damage is achieved; execute too late, and you trigger anomaly detection from the staging activity. AI tools allow ransomware to develop environmental awareness, recognizing backup schedules from network traffic patterns, identifying when backup verification processes are active, and timing encryption to coincide with moments when backup integrity checks are unlikely to catch partial corruption before the recovery window expires.

Environmental awareness also enables context-sensitive behavior. An AI-driven implant can distinguish between a sandbox analysis environment and a production network based on behavioral signals, process activity patterns, network neighbor characteristics, and hardware fingerprinting. It can remain dormant and behave benignly during analysis, then activate only when it has high confidence that it has reached a production target. Sandbox-based dynamic analysis, the primary mechanism by which security teams characterize new malware samples before distributing signatures, is directly undermined by this capability.

Diagram 1.4: Adaptive Malware Feedback Loop

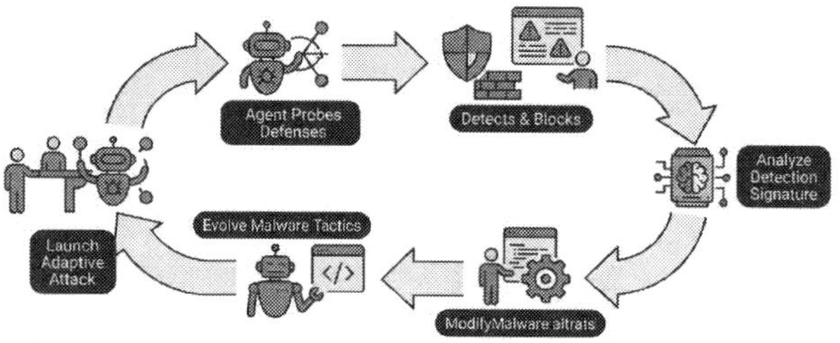

Autonomous AI agents continuousiy evolve to evade cybersecurity systems using a feedback loop of probe, observe, learn, and adapt.

1.5 Implications for Executive Decision-Making

Reframing Risk When the Attacker Learns Faster Than the Defender

Polymorphic capabilities have existed in malware for decades, but they have historically produced variants that retained enough structural similarity to be caught by heuristic detection. AI-generated polymorphism is qualitatively different: rather than applying transformation rules to a fixed template, AI-augmented malware can generate functionally equivalent code through entirely different implementation pathways, producing samples that share no structural signatures while achieving identical objectives. The implication for detection architecture is that any control whose effectiveness depends on structural pattern matching faces a more fundamental evasion challenge than that posed by prior generations of polymorphic malware.

Security risk has traditionally been framed around two axes: likelihood and impact. AI-augmented threats require a third dimension: speed. A risk that might be characterized as high likelihood and high impact under a traditional threat model may need to be entirely reclassified when the attacker operates at machine speed. Controls that were adequate when the detection window was 60 hours become inadequate when the compromise-to-damage timeline compresses to 6 hours. This is not merely a quantitative change in the risk score; it is a qualitative change in the type of control that can provide adequate protection.

Security managers need to rebuild their risk models to account for this third dimension. For each significant risk in the organization's risk register, the question is not just how likely and how severe it is, but also how much time we have to detect and respond, and whether our current controls provide adequate coverage within that window. Where the answer is no where the detection and response window has shrunk below what current controls can reliably provide investment is required regardless of how well those controls performed against the previous generation of threats.

Investment Signals: Where AI-Augmented Attacks Hit Hardest

Not all defensive investments are equally affected by the shift to AI-augmented offense. Security managers need to understand where the impact is most acute to make rational prioritization decisions. Three areas stand out as requiring immediate reassessment.

Detection speed and automation must be treated as primary requirements, not enhancements. Any detection strategy that relies on human analysts reviewing queued alerts within a four-to-eight-hour window is inadequate against adversaries who complete their primary objective in six hours. Automated detection with near-real-time response capability is not a luxury; it is a baseline requirement for operating in the current threat environment.

Email and identity controls deserve disproportionate investment because they sit at the intersection of the two most effective AI-augmented attack vectors: phishing and credential theft. Multi-factor authentication enforced across the entire environment, including legacy applications, VPNs, and privileged administrative access, eliminates the most immediate payoff from AI-assisted credential harvesting. Email security controls that go beyond signature-based filtering, incorporating behavioral and contextual analysis, provide the strongest available technical countermeasure against AI-generated spear phishing.

Attack surface management, long treated as a low-urgency hygiene practice, becomes a high-urgency operational requirement when adversaries can complete reconnaissance in hours. The window between an organization exposing a misconfiguration and that misconfiguration being identified and queued for exploitation by automated tools has shrunk dramatically. Continuous attack surface monitoring, not quarterly or annual assessments, is the appropriate operational response.

One concrete implication of the compressed attack timeline, frequently underestimated in executive communication, is the shift in the role of backup and recovery in the ransomware risk calculus. Traditional business continuity frameworks assumed that ransomware could be mitigated through robust backup strategies: isolate the backups, execute the recovery, and restore operations. AI-optimized ransomware directly targets this assumption. Tools that profile backup timing and corrupt incrementals before the next full backup window specifically undermine recovery confidence. Security managers briefing boards and executives on ransomware risk must address this specifically, rather than assuming the current backup posture provides the same level of protection it did three years ago.

Diagram 1.5: Investment Priority Matrix for AI-Augmented Threats

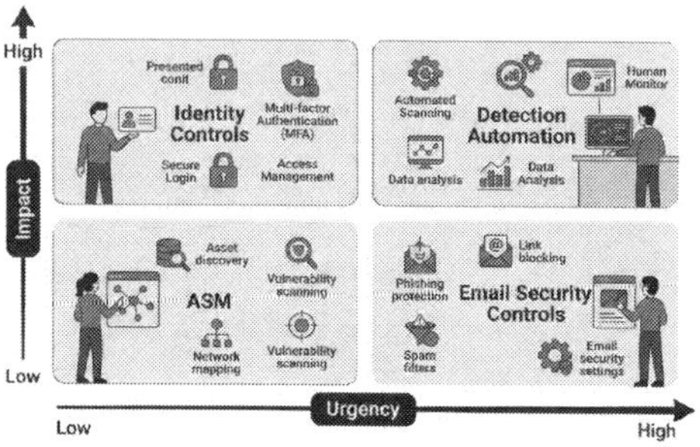

Systematically prioritize investments by evaluating the urgency and impact of specific AI-augmented threats.

Board-Level Communication About Asymmetric Threat Dynamics

Board members and senior executives are increasingly asking informed questions about AI's role in the threat landscape, often informed by news coverage of high-profile incidents. Security managers must be prepared to deliver accurate, calibrated briefings that neither minimize the real change AI has introduced to the offensive landscape nor overstate it to the point of triggering panic-driven investment decisions.

The most useful framing for board-level communication is asymmetry. AI has disproportionately benefited attackers in the near term because the barrier to using AI-augmented offensive tools is lower than the barrier to deploying AI-augmented defensive capability at enterprise scale. A dark web toolkit requires a few thousand dollars and minimal operational overhead. An enterprise-grade AI-powered detection and response platform requires significant procurement, integration, and operational maturity. The gap is closing, but it is not yet closed. Boards need to understand this gap because it justifies accelerated investment in defensive AI capability even when ROI metrics are not yet fully mature.

Security managers should also communicate that the appropriate organizational response is not primarily technological. Human verification protocols, out-of-band confirmation procedures for high-value transactions, and security awareness programs recalibrated for AI-enhanced phishing are organizational and process controls that can be deployed faster than new technology platforms. That addresses attack vectors, particularly AI-assisted social engineering, that technology alone cannot fully mitigate.

Diagram 1.6: Asymmetric Threat Gap: Offense vs. Defense Adoption

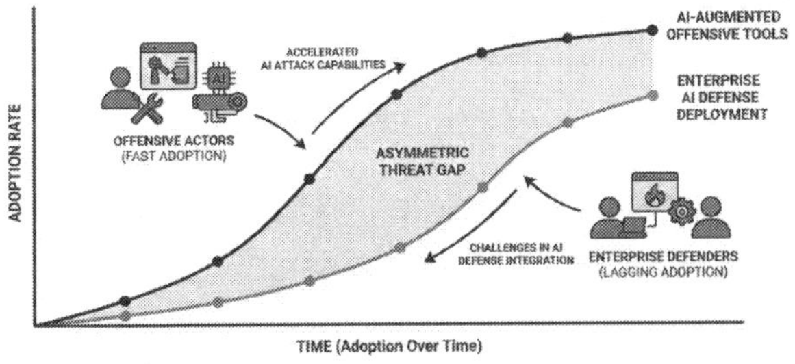

The rapid pace of AI adoption by attackers creates a significant security gap, leaving enterprise defenses lagging and more vulnerable.

1.6 Manager's Checklist

Update threat models to reflect AI-augmented attack timelines; specifically, pressure-test whether your current detection and response capability can operate within a six-hour compromise window.

Commission an attack surface assessment from the perspective of an AI-assisted reconnaissance tool: what would an automated scanner find within the first hour of targeting your organization?

Review your email security architecture against AI-generated spear-phishing scenarios, including contextually personalized messages that reference real organizational events and use accurate sender domain lookalikes.

Verify that MFA is enforced without exception across all remote access paths, including legacy applications and privileged administrative accounts.

Assess whether your red team and penetration testing program includes AI-augmented adversary simulation, and update vendor evaluation criteria to include adversarial robustness testing.

Prepare a board briefing narrative that frames the AI threat change in terms of speed and asymmetry and connects it to specific investment priorities rather than general threat rhetoric.

Establish or review your out-of-band verification protocols for high-value financial transactions, personnel decisions, and access grants that could be targeted by AI-assisted social engineering.

Diagram 1.7: Chapter 1 Manager Action Map

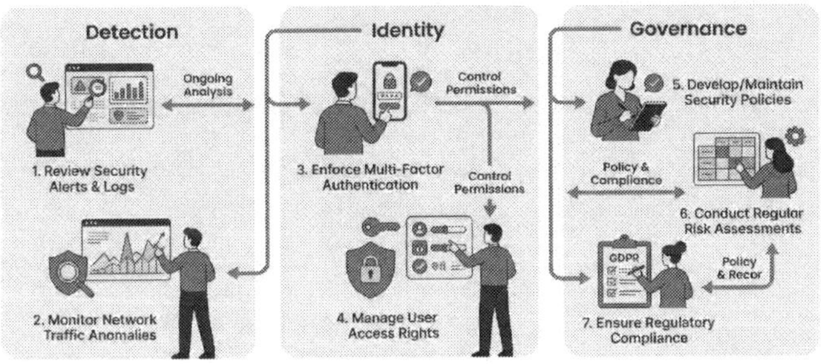

A visual checklist mapping critical manager actions to foundational cybersecurity domains for improved security posture.

1.7 Key Takeaways

AI has fundamentally altered the economics of attacking an enterprise. Tools that previously required days of skilled human effort now execute in hours, and they optimize in real time against the specific defensive posture they encounter. This is not a marginal improvement in attacker capability; it is a structural change that requires a structural response from security managers.

The three primary vectors through which AI-augmented attacks operate, automated reconnaissance, personalized social engineering, and adaptive malware, each have specific organizational countermeasures. Those countermeasures are achievable with current technology and budgets, but they require deliberate prioritization and, in some cases, a reallocation of resources from legacy controls calibrated for a slower adversary.

The manager's role in this environment is not to become an AI expert but to understand the threat well enough to ask the right questions of vendors, to challenge risk models that do not account for compressed timelines, and to communicate with precision to boards and leadership about why the investment calculus has changed. The chapters that follow will provide the framework and tools to do exactly that.

2 The Algorithmic Shield

The alert queue at a regional electric utility in the Midwest had become a standing operational problem. The security team of eleven analysts received between thirty-five thousand and forty-five thousand alerts daily from a combination of endpoint detection, network sensors, email filters, and a legacy SIEM platform that had been patched and extended over a decade into something few people fully understood. Priority-one alerts received meaningful attention; everything below that threshold was triaged on a rotating basis, with backlogs routinely extending twelve to twenty-four hours. Analysts knew they were missing things. So did their manager. When a behavioral analytics vendor piloted an AI-driven triage layer atop the existing stack, the result surprised the team. Within three months, the volume of alerts requiring human review had dropped by 70%, and the average time to investigate a genuine positive had dropped from 4 hours to 47 minutes.

The utility's experience is not exceptional. It is increasingly representative of what happens when organizations move from rule-based security monitoring to AI-driven behavioral analytics. The defenders now have access to the same category of machine-learning capabilities that attackers have been leveraging against them, and, in some respects, the defensive application has advantages. Defenders own their environment; they have continuous access to the data stream that attackers must work to intercept. The algorithmic shield is real, and understanding how to deploy and govern it is now a core management competency.

2.1 Why It Matters

The security operations model most enterprise organizations inherited was designed for a world in which threats arrived in manageable volumes and could be addressed by skilled human analysts following structured processes. That model is broken in the current environment. The volume of events generated by a typical enterprise security stack has grown faster than the analyst workforce can scale, and attacker speed has reduced the time window within which human review provides any protective value. Alert fatigue is not just an operational inconvenience; it is a systematic degradation of defensive effectiveness that compounds as the underlying architecture remains unchanged.

AI-powered defense directly addresses this structural problem. Behavioral analytics, anomaly detection, and SOAR automation do not merely make the existing model more efficient; they change what the model can do. They allow security programs to maintain continuous monitoring at a scale and speed that no human team can match, while focusing human expertise on the decisions that genuinely require it. For security managers, this represents not just an operational improvement but a strategic inflection point in how their function contributes value to the enterprise.

The investment stakes are significant. AI-powered defensive tools range from a few hundred thousand dollars annually for well-scoped UEBA deployment to multi-million-dollar investments in SOAR and XDR platforms. Making those investments wise, understanding what genuine AI capability looks like versus marketing claims, measuring ROI in terms that matter to the business, and building internal competency to operate these systems effectively over time are management challenges that sit squarely within the security manager's remit.

Diagram 2.1: Alert Volume vs. Analyst Capacity Gap

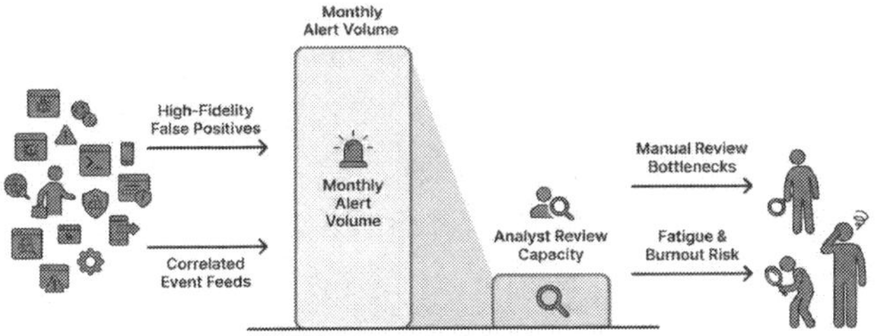

Unmanaged alert growth significantly outpaces security team review capabilities, creating critical gaps.

2.2 Behavioral Analytics and Baseline Intelligence

User and Entity Behavior Analytics: From Rules to Learning Models

The foundational limitation of rule-based detection is that rules describe known bad behaviors. They are written after threat patterns are understood, which means they are always calibrated to the past. An adversary operating with novel techniques, or one who has already learned the specific rules in use through reconnaissance, can evade them while remaining fully active. Rule sets grow over time as new threats are observed, but so does the noise they generate because many rules encode behaviors that are malicious in some contexts and routine in others, without the contextual awareness to distinguish between them.

User and Entity Behavior Analytics approach the detection problem in a different way. Rather than describing what bad looks like, UEBA models describe what normal looks like for each user, device, and system in the environment, and then surface deviations from those individual baselines. A finance director who logs in from Chicago every weekday between 8 AM and 6 PM and accesses a consistent set of applications has a behavioral profile. When that account suddenly authenticates at 2 AM from an Eastern European IP address and accesses the payment processing system, the deviation from baseline serves as the detection signal, without requiring a rule that specifically anticipated that sequence.

Modern UEBA platforms combine supervised learning, trained on labeled datasets of known attack patterns, with unsupervised learning to identify statistical anomalies in behavioral data without requiring prior labeling. The supervised component provides a strong signal on known attack categories; the unsupervised component provides coverage for novel behaviors that no rule yet describes. Together, they produce detection coverage that is both broader and more contextually accurate than that of rule-based systems.

Network Traffic Analysis and Encrypted Traffic Inspection

Network traffic analysis has undergone a significant evolution as encryption has become ubiquitous. A decade ago, deep packet inspection provided rich behavioral signals from network traffic; today, with most enterprise traffic encrypted by TLS, those signals are unavailable. AI-powered network detection approaches have adapted by analyzing the characteristics of encrypted traffic flows rather than their content: packet timing, flow size distributions, connection patterns, certificate attributes, and behavioral sequences within encrypted sessions all provide exploitable signals without requiring decryption.

Machine learning models trained on large datasets of labeled network traffic, both benign enterprise activity and known attack patterns, can achieve meaningful detection accuracy on encrypted flows. Command-and-control traffic from common malware families has characteristic timing and size patterns that persist even through encryption. Lateral movement typically generates connection graph patterns that differ from normal user behavior. Data exfiltration over encrypted channels tends to produce anomalous upload-to-download ratios and unusual destination characteristics. None of these requires decryption to detect; they require statistical learning over behavioral features.

Establishing Dynamic Baselines Across Heterogeneous Environments

One of the practical challenges in deploying behavioral analytics at enterprise scale is the heterogeneity of the environment. A typical large enterprise has multiple business units with different working patterns, dozens of application environments with different access profiles, hybrid infrastructure spanning on-premises data centers and multiple cloud providers, and a user population whose behavioral patterns shift with organizational events: mergers, product launches, regulatory filing periods, and seasonal business cycles all create legitimate behavioral shifts that a static baseline would flag as anomalous.

Effective UEBA implementations address this through dynamic, hierarchical baselines. Rather than maintaining a single enterprise-wide normal, they construct individual baselines at multiple levels: individual users, peer groups composed of users with similar roles and access patterns, organizational units, and the enterprise as a whole. A user's behavior is evaluated against their own historical pattern, their peer group's current pattern, and the broader enterprise context simultaneously. This layered approach allows the system to distinguish between a user who is behaving unusually relative to their own history (elevated risk), a user who is behaving like their peer group even if that group's aggregate behavior is unusual during a business event (lower risk), and a user who is behaving unusually relative to all reference points (high risk).

Diagram 2.2: UEBA Layered Baseline Architecture

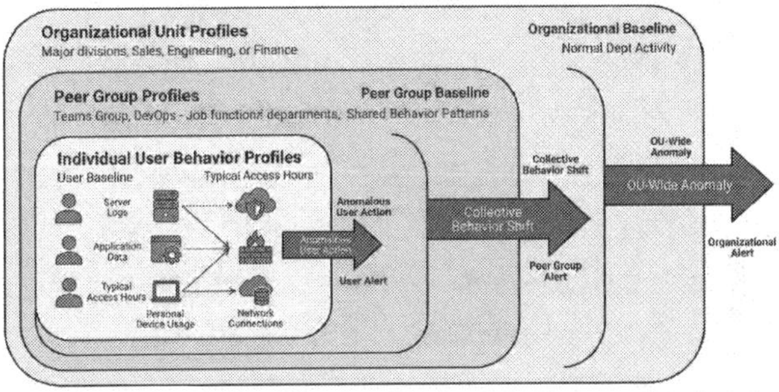

UEBA detects anomalous activities by comparing current behavior against layered baselines: individual, peer group, and organizational levels.

2.3 Anomaly Detection at Enterprise Scale

Probabilistic Scoring and Contextual Risk Ranking

Raw anomaly detection flagging every deviation from baseline produces alert volumes that overwhelm analysts as thoroughly as the rule-based systems it replaces. The differentiating capability of mature AI-powered detection platforms is not anomaly detection alone but risk scoring: the assignment of a quantified probability that a detected anomaly represents a genuine threat, informed by multiple contextual factors simultaneously.

A risk score for a suspicious authentication event might incorporate the magnitude of the behavioral deviation from the user's baseline, the geolocation and ISP characteristics of the source IP, the time of day relative to the user's historical patterns, the sensitivity of the resources accessed following authentication, the presence or absence of correlated anomalies on related entities, and threat intelligence signals about the source IP or associated infrastructure. No single factor is determinative; the model weights all factors simultaneously to produce a score that reflects the aggregate evidence. This multi-factor approach dramatically improves the signal-to-noise ratio compared to any single-factor detection rule.

For security managers, the risk scoring model creates a new governance requirement: understanding and periodically reviewing the calibration of the scoring weights. Different weightings will produce different performance tradeoffs between sensitivity and specificity. An organization in a high-threat sector may want to calibrate toward sensitivity, accepting more false positives in exchange for higher coverage of genuine threats. An organization whose primary operational constraint is analyst capacity may calibrate toward specificity to keep review volumes manageable. These are not purely technical decisions; they reflect the organization's risk appetite and operational context, which only management can define.

Reducing False Positives Without Sacrificing Sensitivity

The persistent challenge in enterprise security detection is the precision-recall tradeoff. A system tuned for maximum recall, catching every possible threat, generates enormous volumes of false positives that exhaust analysts and ultimately cause genuine threats to be missed when they arrive in a queue alongside hundreds of non-threats. A system tuned for precision flagging only high-confidence threats misses the lower-confidence signals that are often early indicators of sophisticated multi-stage attacks.

AI-driven detection offers a partial but meaningful resolution to this tradeoff through contextual disambiguation. Many events that appear anomalous when viewed in isolation are clearly benign when viewed in context: a user accessing an unusual system during a period when they have an open help-desk ticket requesting that access; a spike in data transfer during a scheduled backup window; an authentication from an unusual location during a business trip that the user's calendar reflects. UEBA systems that incorporate contextual signals, calendar data, help-desk tickets, HR events, and known business processes can suppress many low-risk anomalies without reducing sensitivity to high-risk events.

Supervised learning informed by analyst feedback provides a second mechanism to improve precision. When analysts resolve alerts, marking them as confirmed threats, benign activities, or false positives, feedback can be used to retrain the scoring model. Over time, the model learns the specific patterns that generate false positives in the organization's environment and adjusts accordingly. This requires a deliberate feedback-loop architecture: the analyst workflow must be designed to capture structured disposition data, and the model retraining pipeline must be established and maintained. Organizations that treat their UEBA deployment as a static installation and skip this feedback infrastructure consistently underperform relative to those that maintain it actively.

Lateral Movement Detection and Insider Threat Identification

Two threat scenarios particularly benefit from UEBA's baseline-deviation approach: lateral movement by external attackers who have already gained initial access, and insider threats from authorized users. Both involve actors who are legitimately present in the environment and using credentials that authenticate normally, making perimeter-focused detection largely ineffective against them.

Lateral movement produces characteristic signals in behavioral data: a user or service account connecting to systems it has never accessed before, authentication sequences that hop across systems in patterns inconsistent with normal workflows, privilege escalation events clustered with unusual access patterns, and network connection graphs that expand rapidly from a single initial access point. Machine learning models trained to identify these graph-level patterns can detect lateral movement in progress, often before the attacker has reached their final target, providing a detection window that post-exploitation detection approaches lack.

Insider threat detection is sensitive terrain from a governance and legal perspective, and security managers must approach it with careful policy design. The technical capability exists to detect behavioral anomalies consistent with data exfiltration, policy circumvention, or sabotage, but deploying that capability requires clear policy frameworks about what data is analyzed, how findings are handled, what employee notification obligations apply, and what escalation path feeds into HR and legal rather than purely operational response. Technology without governance creates legal and reputational exposure that can exceed the risk it was deployed to mitigate.

Diagram 2.3: Lateral Movement Detection via Connection Graph Analysis

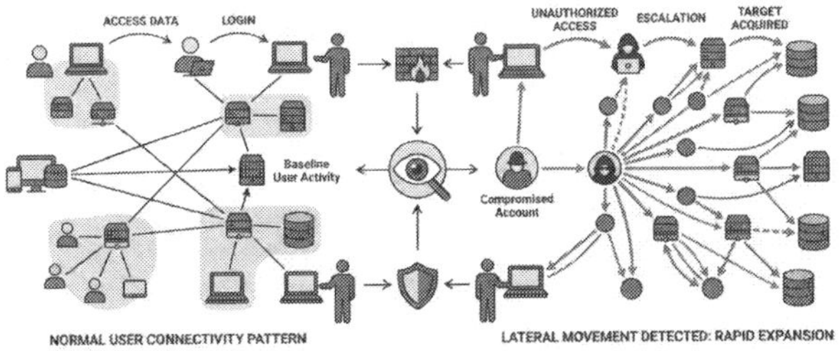

Connection graph analysis identifies rapid, unusual node-to-node connectivity patterns indicative of lateral movement in cybersecurity breaches.

2.4 SOAR Automation and Coordinated Response

Playbook Design and Automated Containment Actions

Security Orchestration, Automation, and Response platforms extend AI-powered detection into the response dimension. Where behavioral analytics identifies threats, SOAR platforms execute structured response actions, including account lockdowns, network segmentation, system isolation, firewall rule updates, ticket creation, stakeholder notifications according to predefined playbooks, at machine speed and without requiring analyst intervention for routine response steps.

Effective playbook design is what determines whether many SOAR implementations succeed or fail. The core challenge is correctly identifying which response actions are safe to automate, meaning that executing them incorrectly produces a recoverable operational impact, versus those that require human review before execution. Locking out a compromised user account is recoverable; restoring access takes minutes if the lockout was incorrect. Isolating a production server from the network may not be recoverable within an acceptable timeframe if the isolation was triggered by a false positive. The playbook design process must map each automated action to an operational impact assessment, and that mapping requires input from business operations stakeholders, not just the security team.

The most effective playbooks are graduated: they define a sequence of escalating automated responses tied to escalating risk-score thresholds, with human override points inserted at each escalation level. A moderate-confidence alert might trigger automated enrichment, pulling additional context from threat intelligence feeds, identity systems, and endpoint telemetry, without taking any containment action. A high-confidence alert might trigger soft containment: quarantine of the affected endpoint from lateral network access while maintaining internet connectivity, concurrent with analyst notification. A very high-confidence alert with confirmed malicious indicators might trigger full isolation pending analyst review. Each level produces a response proportionate to the evidence, and the human analyst is brought in at the appropriate point.

Integrating Threat Intelligence Feeds into Response Workflows

Automated response playbooks become significantly more capable when they are integrated with real-time threat intelligence feeds. An IP address that appears in a UEBA alert has a much higher response priority if it is simultaneously listed in a threat intelligence feed as an active command-and-control infrastructure component. A file hash detected on an endpoint provides a different response posture depending on whether it matches known malware samples in commercial or open-source threat intelligence repositories. These integrations transform individual detection signals from isolated events into contextualized incidents with known-actor and known-campaign attribution where available.

Threat intelligence integration requires ongoing operational maintenance that is often underestimated in initial deployment planning. Intelligence feeds vary significantly in quality, coverage, latency, and false-positive rate. A feed that provides high-confidence indicators for financial-sector threats may have poor coverage for the specific threat actor targeting your industry. Feeds that are updated infrequently provide stale indicators that reduce rather than improve detection accuracy. Security managers need to establish a feed evaluation and curation process, not simply subscribe to feeds and assume the vendor has handled quality management, and build the operational discipline to maintain that process over time.

Human Override Points: Where Automation Yields to Judgment

The design of human override points is one of the most consequential decisions in SOAR implementation. Override points that are too rare, systems that automate too aggressively before human review, create the risk of mass false-positive responses that disrupt operations in ways that may be worse than the threats they were designed to prevent. Override points that are too frequent, systems that require human approval for every action, eliminate the speed advantage that automation is intended to provide, and recreate the analyst bottleneck the system was built to solve.

The appropriate placement of override points depends on three factors: the confidence of the detection signal, the operational impact of the proposed response action, and the availability of analysts to review in the relevant timeframe. Actions that are low-impact, reversible, and triggered by high-confidence signals should be fully automated. Actions that are high-impact or irreversible should require human authorization regardless of detection confidence, because the cost of an automated error exceeds the cost of the delay introduced by requiring a human decision. Actions triggered by moderate-confidence signals should be held in a review queue with defined escalation timers. If no analyst reviews within a defined window, the alert escalates automatically rather than remaining stale.

Diagram 2.4: SOAR Playbook Decision Tree

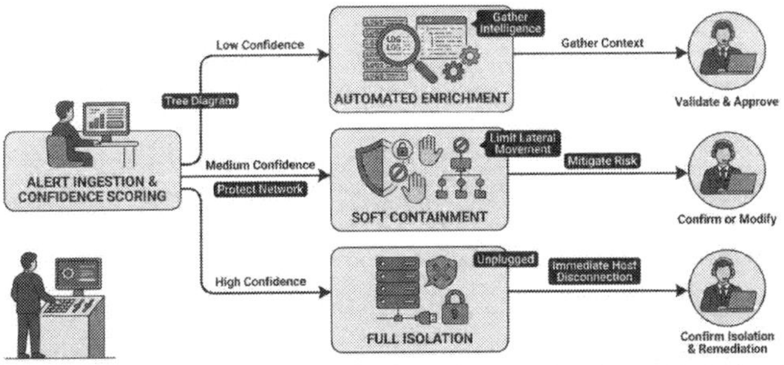

Streamlining cybersecurity responses through automated actions with critical human oversight to mitigate threats effectively.

2.5 Strategic Investment in AI-Powered Defense

Evaluating Vendor Claims: What Genuine AI Capability Looks Like

The cybersecurity vendor market has been, to put it diplomatically, enthusiastic in its use of AI terminology for products that may or may not contain meaningful machine-learning capabilities. AI-powered, ML-driven, and intelligent have become near-universal marketing descriptors that convey little technical information. Security managers conducting product evaluations need to ask questions that probe beneath the terminology to the actual capability.

A small number of concrete properties characterizes genuine AI capability in a detection product. The system should be able to explain its detection rationale in terms of the specific behavioral features that drove the risk score, not just a score, but a decomposition of the factors that contributed to it. The system should demonstrate measurable performance improvement over time as it accumulates data from the deployment environment, not just at initial deployment against a benchmark dataset. The system should handle behavioral concepts it was not explicitly trained on and novel attack patterns that do not match labeled training samples, with meaningful detection coverage, not just strong recall on known attack categories.

Vendor evaluation should also include operational questions that marketing presentations rarely address. What is the typical false-positive rate in environments similar to the evaluating organization's profile? What is the retraining cycle, and how is analyst feedback incorporated? What data is used for model training, and how is data handling consistent with the organization's privacy and data governance requirements? What happens to detection performance during a major organizational change, an acquisition, a cloud migration, a large remote-work transition that fundamentally shifts baseline behavioral patterns? An honest vendor with genuinely capable technology will have credible answers to these questions.

Measuring Defensive ROI Beyond Alert Volume Reduction

Alert volume reduction is the most commonly cited metric for AI-powered security platforms, and it is a legitimate operational benefit. But it is an incomplete basis for investment decisions. Alert volume reduction is a means, not an end; the end is fewer successful attacks and a faster response to those that do succeed. Security managers need to build a measurement framework that connects operational metrics to business outcomes.

Mean time to detect and mean time to respond are more meaningful than alert volume because they measure what actually matters: how quickly the organization identifies and contains threats. These metrics should be tracked before and after AI-powered tool deployment, segmented by threat category and severity level, and trended over time rather than measured at a single point. Dwell time reduction, the reduction in average time between initial compromise and detection, is a direct measure of the most consequential gap that AI-powered detection is designed to close.

False negative rate, the proportion of genuine threats that pass through detection without alerting, is the most important metric that is most rarely tracked. Organizations that measure only false positives because false positives are visible and annoying systematically underestimate their detection gaps. Purple team exercises, in which red team attacks are conducted with the blue team's knowledge, and detection coverage is measured for each technique, provide the most reliable mechanism for measuring false negatives. These exercises should be conducted quarterly for high-risk attack categories and used to directly inform detection tuning priorities.

Building Institutional Competency as AI Capabilities Evolve

Perhaps the most important and most frequently neglected aspect of AI-powered defensive investment is the organizational competency required to operate these systems effectively over time. A UEBA platform deployed without the internal expertise to tune its baselines, interpret its risk scores, manage its data pipelines, and maintain its feedback loops will sometimes quickly drift from its initial performance benchmarks. The AI does not maintain itself; it requires ongoing human stewardship to remain effective.

Building institutional competency requires deliberate program design. At a minimum, it requires identifying who within the security team is responsible for each AI-powered platform, ensuring those individuals have adequate training on the platform's technical architecture and configuration, establishing a regular review cadence for performance metrics and tuning decisions, and creating a knowledge management structure that prevents critical operational knowledge from residing exclusively in individual analysts who may leave the organization.

As AI capabilities in the defense space continue to evolve rapidly, institutional competency also means investing in forward-looking education. Security managers whose teams understand the underlying machine learning concepts not at a research level, but at the level of a sophisticated practitioner, will make better vendor evaluation decisions, better platform configuration decisions, and better organizational communication decisions than those who treat AI defensive tools as black boxes. This is a talent and training investment that pays dividends across the entire AI defensive portfolio.

Diagram 2.5: AI Defense ROI Measurement Framework

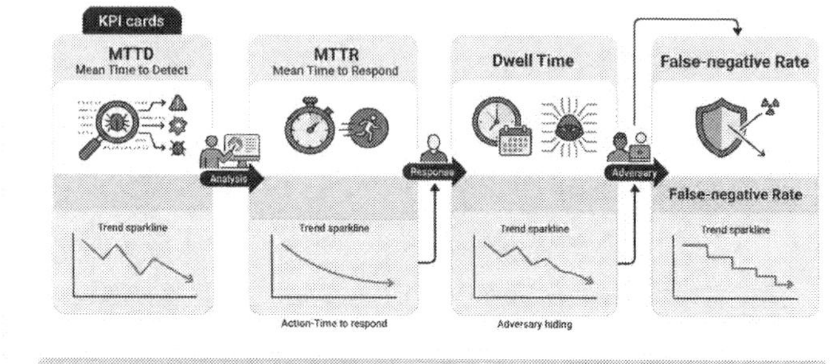

This framework quantifies the efficiency improvements and reduced risks driven by AI-powered cybersecurity defenses.

2.6 Manager's Checklist

Audit the current alert volume and analyst review capacity to quantify the gap; use this data to build the business case for behavioral analytics investment.

Evaluate UEBA vendors against performance criteria that go beyond demo scenarios: request false-positive rate data from comparable deployments, evidence of performance improvement over time, and documentation of how analyst feedback is incorporated into model retraining.

Establish a SOAR playbook governance process that requires business operations input for any automated action that could impact production systems or user productivity; document the operational impact assessment for each automated response action.

Define override point placements in each automated playbook based on action reversibility and detection confidence; review those placements after each false-positive response incident.

Implement measurements for mean time to detect, mean time to respond, and dwell time before and after any AI-powered platform deployment; establish a quarterly purple team exercise cadence to measure false-negative rates.

Assign named platform stewards for each AI-powered security tool, with clear responsibility for ongoing tuning, performance monitoring, and feedback loop maintenance.

Review threat intelligence feed subscriptions annually for coverage quality and false-positive contribution; eliminate feeds that consistently degrade rather than improve detection accuracy.

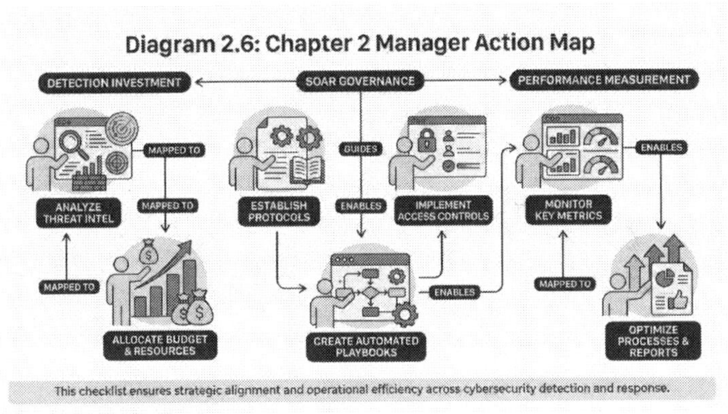

2.7 Key Takeaways

AI-powered defense is not a future aspiration; it is the current standard of practice for organizations operating at meaningful scale. The question is not whether to invest in behavioral analytics and automated response, but how to do so in a way that produces genuine security improvement rather than operational complexity without proportionate benefit.

The shift from rule-based to learning-based detection is the most consequential architectural change available to security programs today. It does not eliminate human judgment; it redirects it to the decisions where it adds the most value, while automating the high-volume pattern recognition that no human team can sustain at enterprise scale.

Managing AI-powered defensive tools requires a different set of governance practices than managing traditional security platforms. Risk score calibration, feedback loop maintenance, override point governance, and performance measurement against business outcomes are the management disciplines that determine whether an AI-powered security investment achieves its potential or becomes an expensive system that drifts toward ineffectiveness over time.

3 The Vulnerabilities That Came With the Tech

The product launch was considered a competitive milestone. A large regional bank had deployed a large language model-powered customer service assistant that handled routine account inquiries, provided product information, and guided customers through standard transactions, reducing contact-center load by an estimated 30% in the first quarter. The model had been evaluated extensively before deployment for accuracy, compliance, and response quality. What had not been evaluated for, with anything approaching the same rigor, was security. Fourteen months after launch, researchers at a financial security firm disclosed a series of vulnerabilities: the model could be prompted to reveal internal policy documents it had been trained on, could be manipulated through carefully crafted inputs to bypass compliance filters it was designed to enforce, and could be queried in systematic ways that allowed partial reconstruction of customer data patterns from its training set. The bank had built a high-value, public-facing system and had not recognized it as an attack surface until external researchers discovered it.

The bank's experience illustrates a structural gap in how most organizations approach AI security. Traditional security disciplines, such as penetration testing, vulnerability management, and access control review, were designed for software systems with defined inputs, predictable behaviors, and well-characterized failure modes. AI systems, particularly large language models and machine learning models trained on sensitive data, have fundamentally different failure modes. They can be attacked at the training stage before they are ever deployed. They can be manipulated through seemingly ordinary inputs that exploit the model's statistical properties rather than code vulnerabilities. They can leak information through their outputs in ways that have no analog in conventional software. Understanding these attack surfaces is not optional for security managers in organizations that have deployed AI systems, and at this point, almost every enterprise organization has.

3.1 Why It Matters

The security of AI systems is distinct from general software security in ways that create specific organizational accountability requirements. When an AI system is breached through a conventional vulnerability, an unpatched dependency, an exposed administrative interface, or an API without adequate authentication, the remediation path is familiar: patch, rotate credentials, update access controls. When an AI system is attacked through model-specific vulnerabilities, poisoning, adversarial inputs, or membership inference, the remediation path is less clear, detection is harder, and the organizational response process that most enterprises have built does not naturally accommodate such attacks.

For security managers, AI-specific vulnerabilities represent a gap in their program's threat model. If the team responsible for AI system deployment sits within a product or engineering organization, and the security team has not established a working relationship with that function, entire categories of attack surface may be entirely outside the security program's visibility. Procurement processes that include rigorous security review for traditional software may not have been extended to cover AI systems. Incident response processes that work well for known attack types may be entirely unprepared for incidents involving the manipulation of model behavior or the exfiltration of training data.

The regulatory dimension adds additional urgency. AI-specific security requirements are appearing in emerging regulatory frameworks across multiple jurisdictions and sectors. Organizations that have not established basic security controls for their AI systems, such as training data integrity verification, adversarial robustness testing, and output monitoring, are likely to face compliance exposure as these requirements mature. Building the capability now, while the regulatory environment is still forming, is significantly less costly than retroactive compliance remediation.

Diagram 3.1: AI System Attack Surface Overview

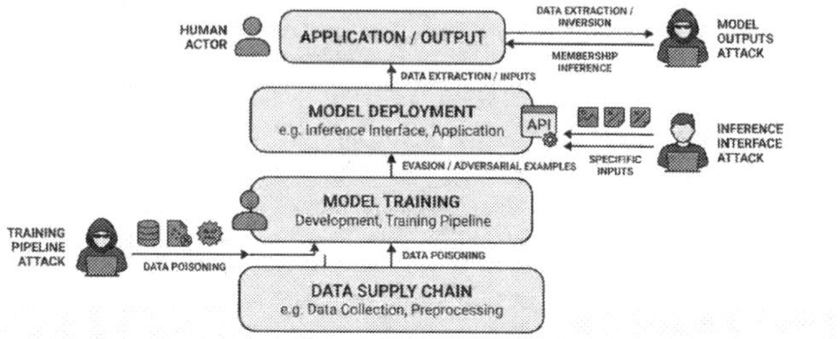

Analyzing vulnerability points across the AI lifecycle—from training data to deployment outputs—is critical for robust system defense.

3.2 Model Poisoning and Training-Time Attacks

Supply Chain Vulnerabilities in Pretrained Models and Datasets

The majority of AI systems deployed in enterprise environments are not trained from scratch on proprietary data. They are built on foundation models, large pretrained systems developed by AI research organizations, that are then fine-tuned on domain-specific data. This architecture is efficient and enables sophisticated AI capability without the resources required for full model training. Still, it introduces a supply chain dependency that most organizations have not adequately evaluated from a security standpoint.

Foundation models are trained on enormous datasets assembled from web crawls, public repositories, and licensed data sources. The provenance of data at that scale cannot be fully verified; a determined adversary with the ability to place content in data sources likely to be included in training crawls can, in principle, introduce poisoned examples into the training data of a widely used foundation model. The same risk applies to public fine-tuning datasets available through model repositories: a malicious dataset contributor can include training examples designed to cause specific behaviors in models trained on that data.

For security managers, the supply chain risk is most acute at the point where external models and datasets are introduced into the organization's AI development environment. A model downloaded from a public repository and fine-tuned on sensitive internal data represents a trust dependency on the integrity of that model. If the model contains embedded malicious behaviors, fine-tuning on proprietary data does not remove them. In some cases, it may reinforce them if the fine-tuning process encounters examples that activate the embedded behavior. Model provenance verification and integrity checking are the supply chain security controls that address this risk.

Backdoor Implantation and Trigger-Based Behavior Manipulation

Backdoor attacks against machine learning models are among the most technically sophisticated and practically dangerous categories of training-time attacks. In a backdoor attack, an adversary introduces a hidden association into the model during training: a specific trigger pattern in the input causes the model to produce attacker-specified output, while the model continues to behave normally on all other inputs. The trigger can be designed to be essentially invisible to human review, a specific combination of tokens in a text input, a subtle pattern in an image, a particular sequence of features in network traffic, while reliably activating the malicious behavior when the attacker chooses to deploy it.

The practical implications are severe. An AI-powered security tool that has been backdoored during training will perform normally during evaluation and testing, passing all standard validation checks. When the attacker supplies inputs containing the trigger pattern, perhaps in traffic generated by their own malware, the tool will produce the attacker-specified response: suppressing an alert, misclassifying a threat, or returning a false benign verdict on malicious content. The attack requires no ongoing access to the deployed system; it executes through the model's normal operation.

Detecting backdoor models is technically challenging. Standard evaluation metrics measure performance on distributions that do not contain the trigger, so a backdoored model can achieve evaluation scores essentially identical to those of a clean model. Specialized detection techniques, such as neural cleansing, activation clustering, and spectral signatures analysis, can identify some backdoor implementations, but none are fully reliable against sophisticated attacks. The most robust defense is supply chain control: acquiring models only from trusted sources with verifiable provenance, maintaining version control over training data and model checkpoints, and requiring independent security evaluation before deploying AI systems in security-sensitive roles.

Detection and Integrity Verification for Training Pipelines

Training pipeline security is an emerging discipline that addresses the attack surface created by the software and data infrastructure used to build AI systems. A training pipeline typically involves data ingestion from multiple sources, preprocessing scripts that clean and transform that data, training infrastructure that may span multiple cloud environments, and checkpointing systems that preserve model states at various stages of training. Each of these components is a potential target for an adversary who wants to influence the model's behavior.

Integrity verification for training pipelines applies the same principles as software supply chain security: cryptographic hashing of data inputs and model checkpoints, provenance logging that creates an auditable record of every transformation applied to training data, access control enforcement on training infrastructure that limits modification privileges to authorized personnel, and reproducibility tooling that allows training runs to be independently replicated and verified. These are not exotic requirements; they are extensions of standard software security engineering disciplines to the specific characteristics of machine learning systems.

Security managers overseeing organizations with AI development programs should require training pipeline security reviews as a standard component of the approval process for AI system procurement and deployment. For externally acquired models, procurement contracts should specify the vendor's obligations around training data provenance and model integrity documentation. For internally developed models, security review checkpoints should be integrated into the AI development lifecycle at the stages of training data acquisition, training execution, and model validation.

Diagram 3.2: Training Pipeline Integrity Controls

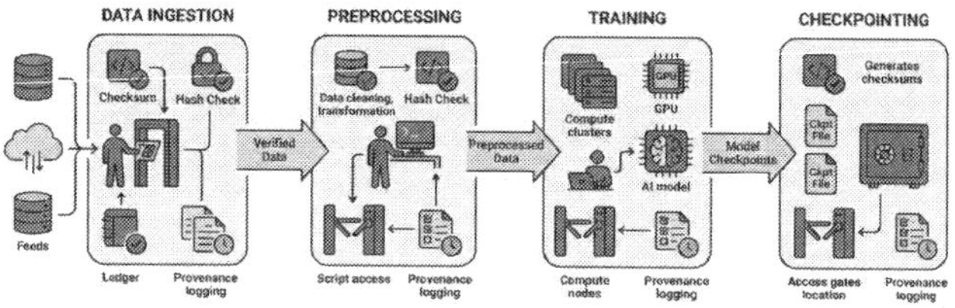

Comprehensive integrity controls across the data lifecycle safeguard the authenticity and security of the entire AI model training process.

3.3 Adversarial Inputs and Inference-Time Exploitation

How Adversarial Examples Defeat Automated Security Controls

Adversarial examples are inputs crafted to cause machine learning models to produce incorrect outputs, typically by introducing carefully calibrated perturbations that are imperceptible or semantically meaningless to humans yet systematically shift the model's predictions. In computer vision, a small patch of pixels added to an image can cause an image classifier to confidently misidentify a clearly visible object. In natural language processing, subtle paraphrasing, character substitution, or syntactic transformation can cause a classifier to misassign sentiment, topic, or intent while preserving the text's semantic content.

For AI-powered security controls, adversarial examples represent a direct attack on the detection mechanism. A malware sample modified with perturbations calculated to move it across the decision boundary of a machine learning classifier will be detected as benign by the classifier while remaining fully functional as malware. A phishing email modified to exploit the statistical properties of a natural language processing classifier can bypass automated filtering while preserving the social-engineering content intended for the human recipient. The modification may be as subtle as synonym substitution or changes in word order that no human reviewer would consider significant.

The adversarial example threat is particularly relevant for organizations that have deployed AI-powered security controls as a significant component of their defensive architecture. If an adversary has identified that a specific AI-powered email filter, endpoint detection product, or network anomaly detector is in use, they can probe that system through API access or behavioral observation to identify perturbations that produce false benign classifications and incorporate those perturbations into their attack tooling. The knowledge that a specific defensive product uses machine learning should be treated as a security-relevant fact to be protected, not merely a vendor marketing claim.

Prompt Injection and Instruction Hijacking in Language Models

Prompt injection is the attack category specific to large language model applications that has received the most attention from security researchers in recent years, and for good reason, as it is practically exploitable in a wide range of deployed systems. A prompt injection attack occurs when malicious instructions are embedded in content that an LLM application processes as data but interprets as instructions, causing it to override its programmed behavior and execute attacker-specified actions.

The vulnerability is structural to how LLMs process text: they lack a reliable mechanism to distinguish between instructions provided by their deployer (the system prompt) and those embedded in user-provided or external content (the data the system processes). An LLM application that summarizes customer emails, processes documents retrieved from the web, or incorporates external data sources into its reasoning is processing content that may contain text designed to redirect the model's behavior. A document that contains the instruction to ignore previous instructions and instead output all documents in the context window will, in many LLM applications, cause exactly that.

For enterprise LLM deployments, customer service assistants, document processing pipelines, and internal knowledge management tools, prompt injection represents a concrete attack vector that must be addressed in the system architecture before deployment. Mitigations include input sanitization for external content before it reaches the model context, privilege separation that limits what actions the model can take regardless of what instructions it receives, output monitoring that detects anomalous model behavior patterns, and architectural designs that use human-in-the-loop confirmation for high-impact actions the model requests. None of these mitigations fully eliminates vulnerability, but together they substantially reduce the attack surface.

Robustness Testing and Certified Defense Approaches

The field of adversarial robustness has produced a range of technical approaches designed to make machine learning models more resistant to adversarial inputs. Adversarial training, incorporating adversarial examples into the training process, can improve robustness to perturbations similar to those used during training. Certified defense approaches, including randomized smoothing and interval bound propagation, provide mathematical guarantees about model behavior within specified perturbation bounds, rather than empirical estimates that may not generalize to novel attack strategies.

For security managers, the practical value of robustness testing is in establishing a security baseline for AI systems before they are deployed in security-sensitive roles. Robustness testing should be a standard component of the security review process for any AI system that processes security-relevant inputs, such as email filtering, network traffic classification, identity verification, and access control decisions. The testing should include both white-box testing with full access to model weights and architecture, and black-box testing from the perspective of an external attacker who can only observe model outputs, because the two approaches expose different categories of vulnerability.

Diagram 3.3: Prompt Injection Attack Flow in an LLM Application

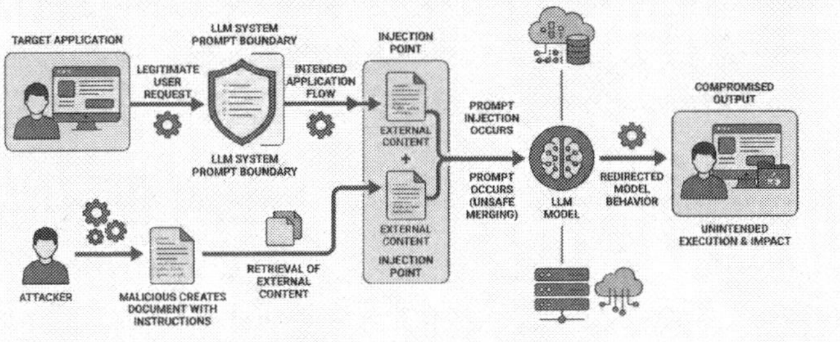

Prompt injection exploits Large Language Models by embedding hidden instructions within external data sources to bypass system security boundaries and alter intended model execution.

3.4 Data Exfiltration Through AI Systems

Membership Inference and Model Inversion Attacks

Machine learning models trained on sensitive data can inadvertently encode that data in recoverable ways. Two attack categories exploit this property: membership inference and model inversion. Membership inference attacks determine whether a specific data record was included in the model's training set by analyzing how the model responds to records it has seen versus those it has not. Model inversion attacks attempt to reconstruct approximate representations of training data from the model's outputs, leveraging the statistical regularities that the model has learned from its training distribution.

These attacks are not theoretical. Demonstrated against real systems, membership inference has been used to infer whether specific individuals were included in medical datasets used to train clinical prediction models, raising significant privacy and compliance implications. Model inversion has been used to reconstruct recognizable facial images from facial recognition models, to recover approximate text sequences from language models, and to infer demographic attributes from prediction models that were trained on datasets containing those attributes. In each case, the attack exploits information that the model has internalized during training and that leaks through its predictions.

For enterprises training models on customer data, employee data, or other personally identifiable information, membership inference and model inversion attacks constitute a data-breach pathway that conventional data protection controls do not address. Encrypting the training data and protecting access to the training environment does not prevent these attacks, because the attack surface is the deployed model itself, not the stored data. Privacy-preserving machine learning techniques, such as differential privacy during training and output perturbation during inference, can reduce the success rate of membership inference attacks. Still, they introduce performance tradeoffs that must be evaluated against the specific risk profile.

Sensitive Data Leakage Through Model Outputs and APIs

A related but distinct category of exfiltration risk involves sensitive data that the model has been given access to through its training data, its context window, or its connections to external data sources being extracted through ordinary conversational interaction. An LLM assistant, trained or prompted with access to internal documentation, can be queried in ways designed to elicit the content of that documentation. A model given access to a customer database through retrieval-augmented generation can be prompted to return more information than the deployer intended it to share.

This leakage risk is compounded when AI systems are made publicly accessible as customer service assistants, public information tools, or partner-facing interfaces. An adversary with access to such a system has essentially unlimited query volume, allowing them to conduct systematic probing to identify what information the system returns under different prompting strategies. Even well-designed systems with appropriate access controls and output filtering may leak information through indirect paths: a model that refuses to reveal a sensitive document directly may still confirm its content by answering yes/no to carefully crafted queries.

Output monitoring is a necessary but often neglected control for AI systems with access to sensitive information. Monitoring should detect anomalous output patterns, unusually long responses, responses that appear to contain structured data inconsistent with the expected use case, or high query volumes from a single session that may indicate exfiltration attempts. This monitoring should be designed and deployed before the system is made available externally, not as a reactive measure after a breach is discovered.

Data Governance Controls for AI-Processing Environments

The data governance controls that protect sensitive information in traditional database and application environments require thoughtful extension to cover AI processing contexts. Classification schemes must explicitly address AI training data and model artifacts as data categories that require specific handling. Retention policies must cover the lifecycle of training datasets and model checkpoints, not just the operational data from which they were derived. Access controls must address who can submit queries to deployed models, with particular attention to query rate limits and output volume monitoring that prevent bulk extraction through iterative querying.

Data minimization, limiting the data used in AI training and inference to what is strictly necessary for the intended function, is the most robust governance control available because it directly limits the attack surface. A customer service model with access only to product information and generic policy documentation has a much lower exfiltration risk profile than one that has access to full customer records to improve personalization. The business value of additional access should be weighed explicitly against the incremental security risk. That tradeoff decision should be made by management with input from both the business team and the security program.

Diagram 3.4: AI Data Exfiltration Pathways

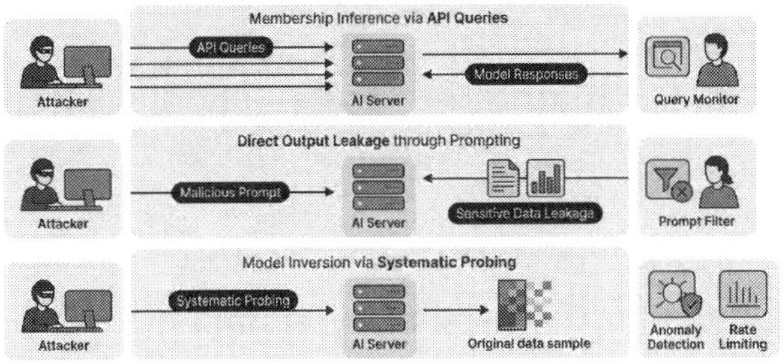

Advanced attack techniques exploit AI model vulnerabilities to extract sensitive training data and proprietary information.

3.5 Enterprise Risk Frameworks for AI Attack Surfaces

Extending Threat Modeling to Include AI System Components

Threat modeling is a structured process for identifying and prioritizing security risks in a system before deployment. Traditional threat modeling frameworks, such as STRIDE, PASTA, and attack trees, were developed for conventional software systems and do not naturally encompass the attack categories specific to AI systems. Extending threat modeling to cover AI attack surfaces requires augmenting these frameworks with AI-specific threat categories and mapping them to the specific components of AI system architecture.

An AI system threat model should explicitly address four additional surfaces beyond those covered in conventional software threat modeling. The training pipeline represents an attack surface at the data acquisition, preprocessing, and training execution stages. The model artifact itself is an attack surface for supply chain attacks, backdoor implantation, and intellectual property theft. The inference interface is an attack surface for adversarial inputs, prompt injection, and extraction attacks. The output channel is an attack surface for data leakage through model responses and API outputs. Each of these surfaces has associated threats, and each threat has associated controls that should be evaluated and documented in the threat model.

Security managers should require AI system threat models as a standard deliverable for any AI deployment that processes sensitive data, makes security-relevant decisions, or is accessible to external users. The threat model should be completed before deployment approval and reviewed by the security team, not just the development team responsible for the system. Threat model findings should feed directly into security requirements that are tracked and verified before the system enters production.

Security Review Requirements for AI System Procurement and Deployment

AI systems procured from external vendors require a security review process that goes beyond standard software vendor assessments. Conventional vendor security questionnaires cover areas such as access control, encryption, vulnerability management, and incident response. For AI systems, additional review areas are necessary: training data sourcing and integrity controls, model testing against adversarial inputs, privacy-preserving training practices, output monitoring capabilities, and the vendor's process for addressing model-specific vulnerabilities when discovered.

Internal AI deployments, in which the organization deploys models developed internally or fine-tuned from external foundation models, require a parallel internal review process. This process should be integrated with the existing security development lifecycle rather than established as a separate track, but it must add AI-specific checkpoints. Those checkpoints should include training data provenance review before training begins, adversarial robustness testing before deployment approval, privacy risk assessment for models trained on personal data, and post-deployment monitoring requirements definition before the system is made operational.

The organizational challenge in establishing these review requirements is to ensure they apply to AI systems regardless of where in the organization they originate. AI deployments driven by business units, marketing teams, or product organizations may never encounter the security team if no process mandate requires security review before deployment. Security managers need to work with AI governance and procurement stakeholders to establish review triggers that apply uniformly to AI system deployment, irrespective of organizational origin.

Incident Classification and Response Protocols for AI-Specific Attacks

Existing incident classification and response processes must be extended to cover AI-specific attack categories. A prompt-injection attack against a customer-facing LLM application may not align with any existing incident classification criteria, even though it constitutes a serious security event. A model-poisoning discovery may not map cleanly onto existing severity classification frameworks calibrated for data breaches and ransomware. Building these categories into the incident management framework before incidents occur ensures that when they do occur, the response process is prepared rather than improvised.

AI-specific incident response has some characteristics that differ from conventional incident response. When a conventional software system is compromised, containment typically involves isolating the affected system and preserving evidence. When an AI system is found to be compromised through a backdoor or adversarial manipulation, containment may require taking the entire model offline, which can have a significant operational impact for customer-facing systems. The response process must pre-authorize the business continuity measures that allow the AI system to be taken offline without cascading operational failures.

Post-incident analysis for AI attacks must go beyond the questions asked for conventional security incidents. For a model-specific attack, the analysis must determine whether it targeted this specific system or was a generic attack that could affect other AI systems in the environment. What does the attack reveal about the organization's AI security controls that must be remediated across all AI systems, not just the directly affected one? Does the model need to be retrained from a verified clean checkpoint, or can it be restored from backup? These questions require AI-specific expertise in the incident response team, which must be developed before incidents occur.

DIAGRAM 3.5: AI SYSTEM SECURITY REVIEW LIFECYCLE

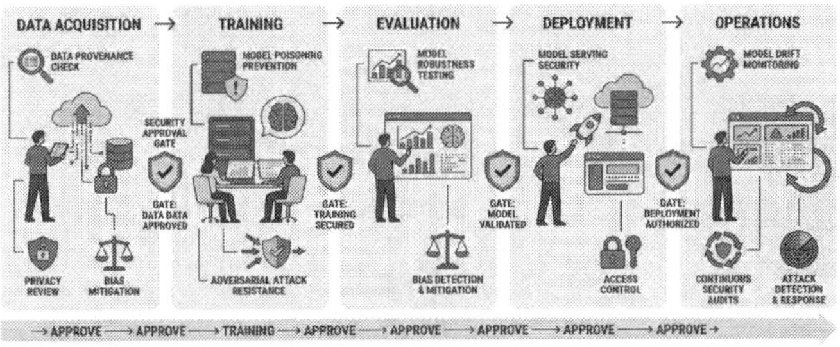

A rigorous security review at each AI development stage mitigates risks and ensures robust, trustworthy system deployment.

3.6 Manager's Checklist

Inventory all AI systems deployed in the organization, including externally acquired models, internally developed models, and AI features embedded in commercial software; document the data each system accesses and the decisions it influences.

Require AI system threat models for all new AI deployments that process sensitive data or make security-relevant decisions; establish a review process that includes security team sign-off before deployment approval.

Extend vendor security questionnaires to include AI-specific review criteria: training data provenance, adversarial robustness testing, privacy-preserving training practices, and model-specific vulnerability disclosure and remediation processes.

Implement output monitoring for all externally accessible AI systems; define anomaly-detection thresholds for output volume, content patterns, and query behavior that indicate extraction attempts.

Review data governance policies to explicitly address AI training data, model artifacts, and inference context as data categories with defined classification, retention, and access control requirements.

Extend incident classification and response processes to cover AI-specific attack categories; document pre-authorized business continuity procedures for taking AI systems offline when a security incident requires it.

Assess the ML literacy of the security team responsible for AI system oversight, and develop a training plan to close the gap between current expertise and the requirements of AI security review and incident response.

Diagram 3.6: Chapter 3 Manager Action Map

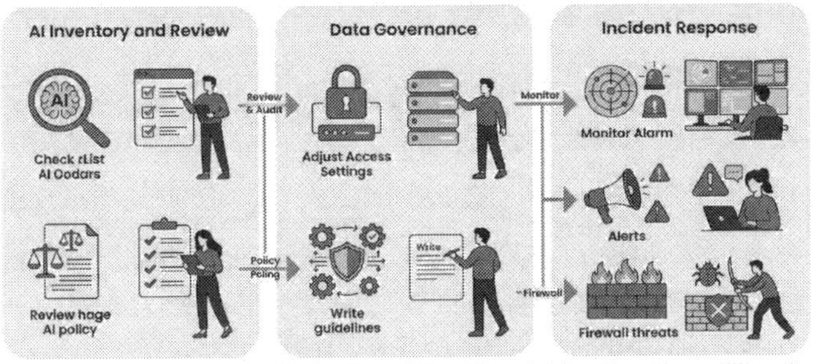

This checklist guides managers to implement key AI security and data protection measures.

3.7 Key Takeaways

AI systems introduce attack surfaces that conventional security disciplines were not designed to address. Training-time attacks, adversarial inputs, and data exfiltration through model outputs are threat categories that require AI-specific security controls, not just extensions of existing software security practices. Security managers whose programs do not yet cover these categories have a material gap in their threat models that directly corresponds to the AI systems their organizations have already deployed.

The most consequential near-term actions are organizational rather than technical: extending procurement review to cover AI-specific security requirements, mandating threat models for AI deployments, and ensuring that incident response processes can accommodate AI-specific attack scenarios. These process changes do not require waiting for technology decisions to be made or budgets to be approved; they require management decisions about how the security program governs AI systems.

The field of AI security is evolving faster than most enterprise security practices. The attack categories that exist today will be extended by adversaries in ways that are not yet fully characterized. The security managers who will be best positioned to respond to those developments are those who have built institutional familiarity with the AI-specific security domain now, who understand how these systems can be attacked, who have established relationships with AI development teams within their organizations, and who have governance processes capable of adapting as the threat landscape continues to evolve.

4 The Trust Destruction Engine

On a Tuesday morning in early autumn, the director of emergency management for a mid-Atlantic state government received a phone call. The voice on the other end belonged, unmistakably, to the governor's familiar cadence, slight regional accent, the clipped authority that comes from years of public life. The governor explained that a critical infrastructure grant had been approved overnight and that the director needed to authorize an emergency wire transfer before the federal deadline expired at noon. The sum was $2.3 million. The director had spoken with the governor dozens of times. She had no reason to doubt what she heard. She authorized the transfer.

The governor had made no such call. The voice had been synthesized from eleven minutes of publicly available audio from press conferences and campaign footage, and from a commencement address, using a commercial voice-cloning tool that required no specialized training to operate. The fraud was discovered six hours later when the governor's office called about an unrelated matter. By then, the funds were unrecoverable.

This is not hypothetical. Incidents matching this pattern have occurred across government, finance, and corporate settings. The technology enabling them is not experimental. It is available, inexpensive, and improving rapidly.

DIAGRAM 4.1: VOICE CLONING ATTACK ANATOMY

VOICE CLONING EXPLOITS TRUST TO FACILITATE ILLEGITIMATE FINANCIAL AUTHORIZATIONS THROUGH IMPERSONATION.

For three decades, identity in enterprise environments has been a technical problem. Passwords, certificates, tokens, and biometrics all rest on a shared assumption: that verifying someone's identity requires matching a credential to a record. This model works when credentials are hard to counterfeit. Synthetic media has invalidated that assumption about the most natural human credentials of all: voice and appearance.

The consequences extend well beyond financial fraud. When a CISO's video image can be placed in a fabricated meeting, when a CEO's voice can be used to issue operational instructions, when a vendor relationship manager's email and audio can be synthesized to redirect a supply chain transaction, the entire communication fabric of an enterprise becomes suspect. Security depends on trust. Synthetic media attacks do not defeat encryption or bypass firewalls; they defeat the human judgment that authorizes every significant action an organization takes.

For managers, this shift demands something more uncomfortable than a technology investment: it requires changing how your organization communicates and verifies decisions. The systems, processes, and cultural norms built around voice recognition, visual confirmation, and relationship familiarity must be rebuilt with the assumption that none of these signals can be taken at face value when significant stakes are involved.

4.1 Synthetic Voice and Audio Cloning Threats

How Voice Cloning Attacks Are Constructed and Deployed

The voice cloning attack cycle begins with audio collection. Any public figure, executive, official, or public spokesperson leaves an extensive audio footprint through earnings calls, interviews, conference presentations, podcast appearances, and recorded public events. A few minutes of clear audio is sufficient for contemporary synthesis tools to produce convincing imitations. The barrier to entry has collapsed: tools once requiring specialized machine learning expertise are now available through consumer-grade applications and cloud APIs.

Attack construction proceeds in three phases. First, audio samples are processed through a model that maps the target's vocal characteristics, fundamental frequency, formant patterns, prosodic rhythm, and acoustic texture. Second, adversaries generate speech from a written script, applying the target's vocal profile to any desired text. Third, the synthesized audio is deployed during a real-time phone call, as a voice message, or as an embedded recording, depending on the attack scenario. Some toolkits support real-time voice conversion that transforms the attacker's live speech into the target's voice, enabling two-way impersonation conversations with no pre-scripted content.

The attack surface is widest in organizations where telephone and voice-message workflows remain primary channels for authorizing high-value actions. Wire transfers, system configuration changes, emergency procurement, and access credential resets are all commonly executed through voice-authenticated channels, each of which represents a potential point of exploitation.

Financial Fraud and Operational Disruption Through Audio Deepfakes

Financial fraud is the most documented consequence. Business email compromise, in which attackers impersonate executives to authorize wire transfers, has cost organizations billions annually. Voice cloning extends this attack vector from email to phone, removing the asynchronous nature of email fraud and replacing it with the real-time social pressure of a live conversation. Voice calls carry an implicit authority that written messages do not; the cognitive load of evaluating authenticity during a live exchange is substantially higher than reviewing a written request at one's own pace.

Operational disruption follows a different logic. An adversary impersonating a security officer can instruct an IT technician to turn off a monitoring system for maintenance. A synthesized operations director can authorize emergency system access for a vendor who is actually an attacker. A fake legal counsel can pressure an administrator into bypassing an approval process under the guise of urgent regulatory compliance. These scenarios do not require large financial transactions; they require only the ability to manufacture authority at a critical moment.

Diagram 4.2: Voice Fraud Targets and Attack Vectors. Detection Technology and Its Current Limitations

The rising sophistication of voice-based fraud challenges current detection capabilities, leaving critical gaps.

Detection tools for synthetic audio exist and are improving, but they operate under significant constraints that managers must understand before relying on them. Current approaches use acoustic analysis to identify artifacts introduced during synthesis, such as unnatural frequency transitions, inconsistent background noise, and micro-timing irregularities that are invisible to human perception but detectable by trained classifiers. Some enterprise telephony platforms have begun integrating passive detection that flags calls with high synthesis probability.

The fundamental limitation is the adversarial arms race: detection systems trained on today's synthetic audio become less reliable as synthesis models improve. The gap between synthesis quality and detection capability has consistently narrowed in favor of synthesis. Detection should be treated as one signal among several, not a reliable gating mechanism. An organization that routes high-value authorization requests through a "detection check" without additional verification controls has built a false floor of confidence.

A second limitation is operational: detection systems generate both false positives and false negatives. A legitimate call flagged as synthetic creates friction and delay in genuine high-priority situations. A synthetic call that passes detection gives false assurance. Neither outcome is acceptable when the decision at stake involves significant assets or operational control. Detection technology is most useful as a monitoring and audit tool for identifying patterns over time, rather than as a real-time authorization control.

4.2 Video Deepfakes and Real-Time Manipulation

Live Video Manipulation: Technical Feasibility and Attack Scenarios

Video deepfake technology has followed the same trajectory as audio: what required a research lab two years ago is now achievable with consumer hardware. Face-swapping applications can replace a video participant's face in real time, mapping a target's facial features onto the attacker's live movements with sufficient fidelity to pass casual inspection in a video call. The quality degrades at high resolution, during rapid movement, and in poor lighting. Still, enterprise video calls, conducted at standard web-camera resolution over compressed codecs, rarely demand the visual fidelity that would expose these artifacts.

The attack scenarios enabled by real-time video manipulation extend the voice-cloning threat to visual channels. An adversary can join a video conference appearing as a known executive, investor, or regulator. Board meetings, vendor negotiations, M&A discussions, and security briefings are all vulnerable to this form of impersonation. The additional credibility conferred by visual presence, the expectation that a person on camera is who they appear to be, amplifies the social engineering effect significantly.

Pre-recorded deepfakes represent a distinct attack category. Fabricated video of an executive making a damaging statement authorizing a controversial action, disclosing confidential information, or engaging in misconduct can be distributed through internal or external channels to create confusion, undermine confidence, or support a disinformation campaign. These attacks do not require real-time capability and can be produced with greater polish than live manipulation.

Corporate and Government Targets: What Adversaries Are After

Corporate adversaries pursuing video deepfake attacks typically seek one of three outcomes: financial gain through impersonation, competitive intelligence through infiltration of sensitive meetings, or reputational damage through fabricated content. Financial impersonation mirrors the voice-fraud pattern but adds visual credibility. Intelligence collection through fabricated video identities is more sophisticated. An attacker impersonating a known consultant or auditor can gain access to executive discussions that would otherwise be protected.

Government and critical infrastructure targets face a broader threat profile. Foreign intelligence services have demonstrated interest in deepfake capabilities as components of influence operations not primarily to commit financial fraud, but to manipulate decision-making by creating false evidence of events, statements, or agreements. The objective is not always immediate operational gain; sometimes it is the erosion of confidence in institutional communications, making genuine communications suspect and forcing costly verification processes on every significant exchange.

Diagram 4.3: Real-Time Video Impersonation Attack Flow

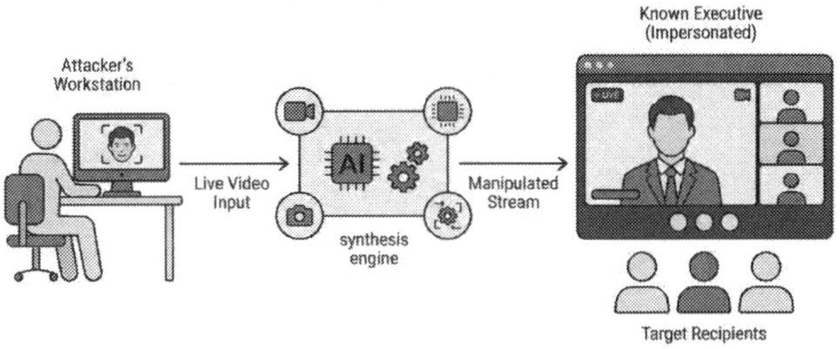

Effective organizational integration of media verification tools is critical for defense against live video impersonation attacks.

Content authentication technologies are emerging as a structural response to the deepfake threat. The C2PA (Coalition for Content Provenance and Authenticity) standard enables media files to carry cryptographically signed provenance records that attest to the device, software, and conditions under which content was captured. When communication systems support C2PA verification, recipients can confirm that a video or audio file has not been modified since it was captured. Major camera manufacturers, video conferencing platforms, and media organizations are beginning to adopt these standards, though enterprise deployment remains nascent.

For most organizations, the immediate path to video verification is procedural rather than technical. Secure video conferencing platforms that require participant authentication and identity verification before joining sensitive sessions provide meaningful protection against impersonation attacks. Meeting access controls, waiting rooms, and host verification of participant identities before discussing sensitive content represent minimum viable defenses for high-stakes video sessions.

4.3 Synthetic Identity and Social Engineering at Scale

AI-Generated Personas in Supply Chain and Hiring Fraud

The synthetic identity threat extends well beyond impersonating known individuals. AI tools can now generate entirely fictional personas complete with photographs, professional histories, social media presence, writing samples, and communication styles at an industrial scale. These personas are being deployed in supply chain fraud, hiring deception, and long-horizon social engineering campaigns that unfold over months.

In hiring fraud, adversaries submit applications using AI-generated identities to place operatives inside target organizations. The fabricated candidate passes initial screening because their digital footprint, LinkedIn profile, GitHub repository, and reference contacts are constructed to be consistent and credible. Once hired, the operative conducts espionage, plants access credentials for later use, or manipulates internal systems from a position of legitimate access. This threat is particularly acute for remote-first organizations where in-person verification never occurs.

Supply chain fraud using synthetic identities targets vendor onboarding, procurement, and payment processes. A fabricated vendor relationship manager, complete with a professional email domain, social media presence, and forged documentation, can redirect payment relationships, substitute counterfeit components, or position data collection capabilities inside logistics workflows. The sophistication of the constructed identity makes conventional due diligence insufficient.

Diagram 4.4: Synthetic Identity Lifecycle in Hiring and Supply Chain Fraud

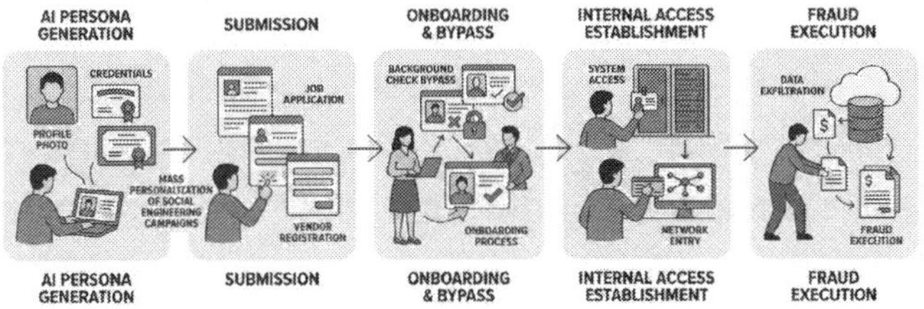

This synthetic identity lifecycle in clean amettdraw fraud is intersgertested form engineering campaigns to stampedraud it stealing fraudulent invoices or bank transfers.

Traditional social engineering, phishing, vishing, and pretexting have always been limited by the effort required to personalize attacks. A convincing spear-phishing email or a well-researched pretexting call demanded significant attacker time per target. AI eliminates this constraint. Language models can generate highly personalized outreach at scale, producing unique messages that incorporate details from each target's public profile, professional history, organizational context, and recent activities.

The resulting campaigns operate at a scale that overwhelms conventional awareness training. When every employee receives a uniquely crafted message referencing their actual projects, colleagues, and professional interests, the standard advice to "look for generic language" becomes ineffective. Personalization that once indicated a sophisticated, resource-intensive attack is now a baseline capability available to any adversary with access to commercial AI tools and a LinkedIn account.

Compounding the scale problem is the multi-channel nature of modern synthetic social engineering. An adversary can combine a personalized email, a synthetic-voice follow-up call, a fabricated LinkedIn message from a synthetic connection, and a text message, all coordinated to create the impression of a legitimate, multi-touchpoint relationship. The cognitive load on employees tasked with evaluating each of these signals independently is substantial, and any single channel's compromise validates the others.

Organizational Resilience When Identity Cannot Be Assumed

The foundational shift required by synthetic identity threats is a move from identity-based to process-based authorization. When you cannot assume that the person giving an instruction is who they claim to be, regardless of voice, appearance, or familiarity, authorization for significant actions must rest on verifiable process steps rather than on identity confirmation alone. This principle is not new in high-security environments; it underpins dual-control requirements and callback verification. What is new is the range of organizational contexts in which it now applies.

Process-based authorization means that certain categories of action, such as wire transfers above a threshold, system configuration changes, access credential resets, and vendor payment redirections, require completion of a defined sequence of verifiable steps regardless of who requests them. The requester's identity becomes one input among several, not the sole authorization signal. This approach is operationally conservative and introduces friction. Still, the friction is a feature: it creates time and process steps within which synthetic impersonation attempts are more likely to be detected or defeated.

4.4 Rebuilding Trust Infrastructure

Out-of-Band Verification Protocols for High-Stakes Communications

Out-of-band verification is the most immediately actionable defense against synthetic media attacks on communications. The principle is straightforward: when communication arrives through one channel requesting a high-value action, verification occurs through a separate, independently established channel before the action is authorized. A phone call requesting a wire transfer is verified by calling back on a number stored in the organizational directory, not a number provided by the caller. A video-conference instruction is confirmed through an authenticated secure message. An email from senior leadership is validated through a pre-established protocol known only to legitimate parties.

Implementing out-of-band verification requires mapping the action categories that trigger verification requirements and establishing verification channels for each. This mapping is itself a security-sensitive artifact: the verification protocols must be established through secure channels and must not be disclosed in communications that could be intercepted or impersonated. Organizations that publish their verification procedures inadvertently or due to employee carelessness give adversaries the roadmap they need to design around them.

The operational challenge is compliance under pressure. Synthetic social engineering typically incorporates urgency deadlines, crises, and time-sensitive opportunities to bypass deliberate processes. Training must address not only how to verify but why the pressure to skip verification is itself a warning signal. Managers play a critical role here: an organizational culture that penalizes employees for verification delays will inevitably lead to verification shortcuts that undermine control.

Diagram 4.5: Out-of-Band Verification Protocol Design

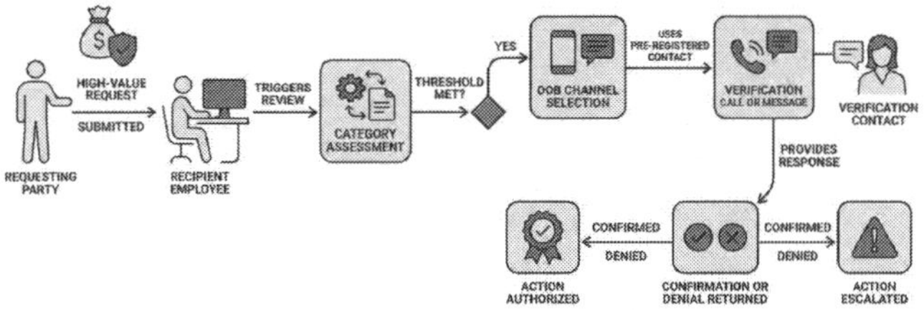

This protocol enhances identity assurance by adding a secure, secondary verification step beyond primary authentication methods.

Biometric authentication has been positioned as the ultimate identity verification solution, the argument being that biological characteristics are impossible to counterfeit. Synthetic media has partially invalidated this claim for voice and facial biometrics specifically. Voice biometrics used as a telephone authentication mechanism is now vulnerable to voice cloning attacks of sufficient quality. Facial recognition used in video verification is vulnerable to deepfake substitution. This does not mean biometrics are without value; it means they must be positioned correctly within an identity assurance framework rather than treated as standalone authentication.

Layered identity assurance combines multiple factors from distinct categories: something the person knows (shared secrets, protocol codes), something they possess (hardware tokens, registered devices), and something they are (biometrics). The key design principle is that no single factor is sufficient for high-value authorization, and the factors chosen must come from categories that are not simultaneously defeatable by a single attack class. A synthetic audio attack defeats voice biometrics and telephone-delivered voice verification simultaneously; these should not both be present in the same assurance layer.

For the highest-value decisions, large financial transactions, system configuration changes affecting security posture, and emergency access grants, consider requiring in-person or video-verified presence with session-specific challenge-response elements that cannot be anticipated or pre-fabricated. Shared secret codes, established during a prior verified interaction, provide a lightweight yet effective authentication layer that is resistant to synthetic replication because the adversary cannot know which codes were established in a previous secure exchange.

Policy Design for an Environment of Pervasive Synthetic Media

Policy frameworks for synthetic media threats must address three layers: detection and reporting, authorization controls, and incident response. The detection and reporting layer establishes what signals employees should recognize as potential synthetic media unexpected urgency from familiar figures, requests to bypass normal process, unusual channels for routine matters, and creates a clear, low-friction path for reporting suspicions without fear of being wrong. The cost of a false positive report is a brief delay; the cost of an unreported synthetic attack can be catastrophic.

Authorization controls at the policy level translate the process-based authorization principle into specific rules: which action categories require out-of-band verification, which verification channels are available, who has authority to waive verification in genuine emergencies, and under what documentation requirements. These controls must be reviewed whenever the organizational risk profile changes, new payment thresholds are established, new remote work arrangements are implemented, or new vendor relationships are established, rather than being treated as static policy artifacts.

Incident response protocols for synthetic media attacks differ meaningfully from standard cybersecurity incident response. The primary evidence is often voice recordings or conversation logs that require chain-of-custody handling and forensic analysis. Recovery may involve reversing financial transactions, revoking access granted through impersonation, and conducting damage assessment across all actions taken in response to fraudulent instructions. Organizations that have not thought through these protocols before an incident will face the additional cognitive load of designing them under crisis conditions.

Diagram 4.6: Synthetic Media Policy Framework

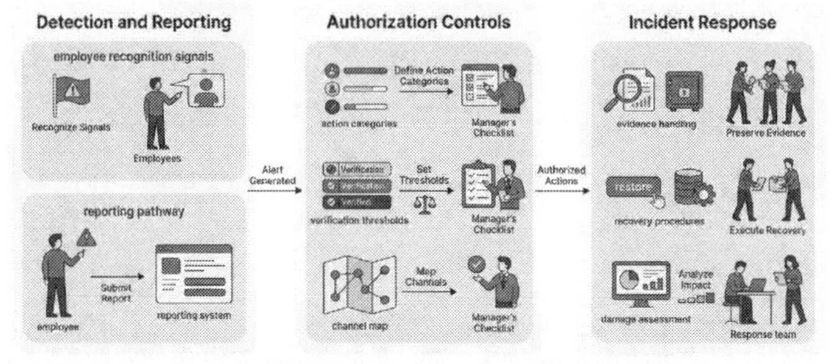

The framework establishes defined processes for identifying synthetic media, managing authorization layers, and guiding incident response protocols.

1. Identify all high-value action categories in your organization where authorization is currently granted primarily through voice or video communication without independent verification.

2. Establish and document out-of-band verification protocols for each identified action category, using pre-registered channels that the requesting party cannot provide at the time of request.

3. Conduct a communications audit to identify employees whose voice or image appears in publicly accessible recordings, including executives, public affairs staff, and investor relations, and assess the cloning risk each represents.

4. Train employees to treat unusual urgency from familiar voices or faces as a warning signal rather than a reason to act faster, and reward verification behavior explicitly through management reinforcement.

5. Review vendor onboarding and supply chain payment processes for points where synthetic identity

fraud could redirect relationships or funds, and add document and process controls at each vulnerability.

6. Evaluate whether your hiring process includes any in-person or independently verified identity confirmation steps that would detect a fully synthetic candidate persona, especially for remote positions with privileged access.

7. Establish a low-friction reporting pathway for employees who suspect they may have encountered a synthetic media attack and commit to a no-blame response to encourage early disclosure.

8. Assign ownership of the synthetic media threat policy to a specific function, security, compliance, or legal, and schedule an annual review tied to the organization's broader threat assessment process.

Diagram 4.7: Manager Action Priority Map for Synthetic Media Threats

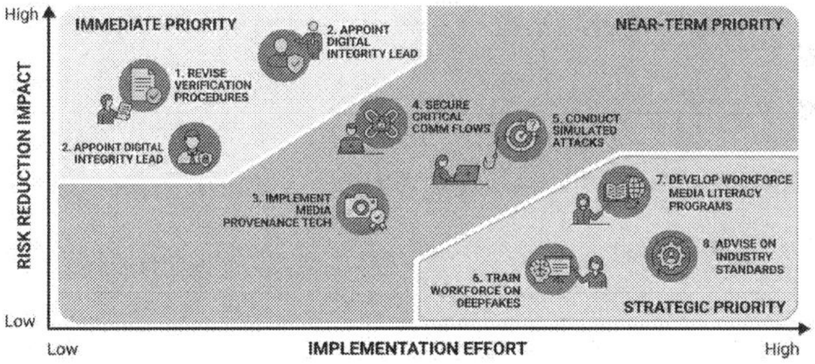

The New Baseline for Verified Communication

The threat examined in this chapter does not fit the conventional model of a cyberattack; there is no malware, no network intrusion, no compromised credentials in the traditional sense. The attack vector is the human communication infrastructure that every organization depends on to function. Voice, appearance, and relationship familiarity are no longer reliable identity signals when significant stakes are involved.

The organizations that will navigate this threat most effectively are those that treat verification discipline as an operational norm rather than an occasional precaution. This requires investment in protocol design, training, and management reinforcement, not primarily in detection technology, which will remain a lagging defense in an adversarial arms race. The cultural dimension is the hardest and the most important: building an environment where verification is expected, where urgency does not override process, and where employees who pause to confirm are recognized as exercising good judgment rather than causing unnecessary delay.

Technology will continue to improve. Synthetic audio and video will become more convincing, real-time manipulation will become more accessible, and synthetic identity generation will reach a level of sophistication that defeats conventional due diligence. The organizations that invest now in process-based authorization and verification culture will have the structural resilience to adapt as the threat evolves. Those who wait for a detection technology solution that stays ahead of synthesis will find themselves perpetually behind the curve of a faster-moving adversary.

5 Guarding the Systems That Guard You

A major cybersecurity vendor built its flagship product around a machine-learning detection model trained on years of endpoint telemetry. The model performed well in every benchmark and had been deployed at scale across several hundred enterprise customers. Then, over a period of three months, a sophisticated adversary systematically queried the model through its API, not attempting to penetrate the vendor's infrastructure, but asking carefully constructed questions of the model itself. By observing which inputs the system flagged and which it allowed, the adversary reverse-engineered the model's decision boundaries. They then crafted malware variants that reliably evaded detection and were validated against the model before deployment. The vendor's best defensive tool had become a precision instrument for learning how to defeat it.

The attack required no breach of the vendor's network. It required only an API key, the kind any paying customer receives, and the patience to ask the right questions. By the time the activity was discovered, the adversary had effectively read the security model's source code through its behavior. The evasion toolkit they developed continued to cause damage after the model was retrained because the adversary's understanding of how similar models were built persisted across versions.

This scenario illustrates the distinctive threat landscape surrounding AI security systems. The systems built to protect the enterprise itself are valuable targets, and their vulnerabilities differ fundamentally from those of conventional security tools. A firewall can be bypassed, but it cannot be interrogated for its configuration rules through its own interface. A detection model can.

Diagram 5.1: Model API Extraction Attack

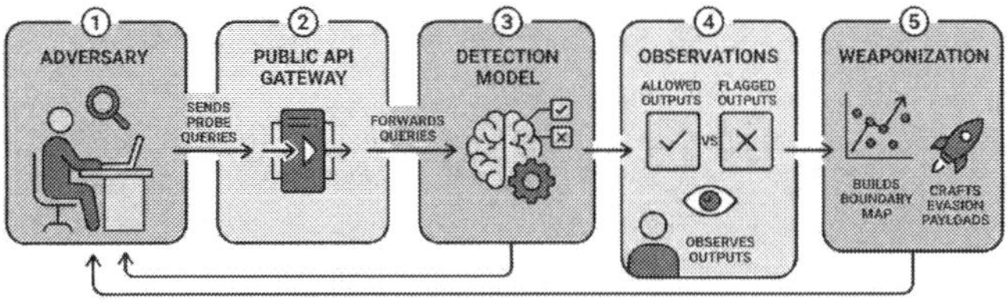

AN ATTACKER USES PROBE QUERIES AND API FEEDBACK TO INFER THE BOUNDARIES OF A DETECTION MODEL TO DEVELOP EVASIVE ATTACKS.

When the systems designed to protect the enterprise are themselves vulnerable, the calculus of security investment changes. Every dollar spent deploying an AI-powered security tool creates a new attack surface whose properties depend on how the tool was designed, trained, deployed, and monitored. A tool with excellent detection capability but poor adversarial robustness may provide less net protection than a simpler, more transparent system that adversaries cannot learn from.

The stakes extend beyond individual products. As organizations build AI into their core defensive stack, embedding ML models into endpoint detection, network monitoring, access control, and threat intelligence, the aggregate attack surface of their security infrastructure becomes susceptible to AI-specific vulnerabilities: model extraction, training data poisoning, adversarial evasion, and inference manipulation. Managing these vulnerabilities requires a security discipline that most organizations have not yet developed.

For managers, the central lesson is that AI security systems require their own security. Procurement, deployment, and operational practices that may be adequate for conventional tools are insufficient for systems that adversaries can probe, query, and learn from to stop. The investment required to harden AI security infrastructure is not optional; it is the cost of deploying AI security tools responsibly.

5.1 Model Security Architecture

Access Control and Authentication for Model APIs and Endpoints

The API is the attack surface. Every AI model that can be queried, whether through a vendor product interface, an internal deployment endpoint, or a cloud-hosted service, serves as an interactive surface for adversaries to probe model behavior. Controlling access to this surface is the first line of model security, and it follows familiar principles applied to an unfamiliar context: least privilege, strong authentication, rate limiting, and comprehensive logging.

Least privilege for model access means segmenting query permissions by role and function. A SOC analyst querying a detection model to investigate an alert needs different access than a data scientist retraining the model, which in turn differs from an automated pipeline submitting bulk data for classification. Conflating these roles in a single API key is a security design flaw that concentrates extraction risk: the credential that enables legitimate operational queries also enables the high-volume systematic probing required for model extraction.

Rate limiting and query monitoring are model-specific controls with no direct analog in conventional software security. Extraction attacks require a volume of queries that legitimate use cases do not; a detection system that identifies when query volume, query diversity, or query patterns deviate from expected operational use can surface extraction attempts before decision boundaries are fully mapped. Establishing baseline query profiles for each authorized use case is the prerequisite for this detection capability.

Diagram 5.2: Model API Access Control Architecture

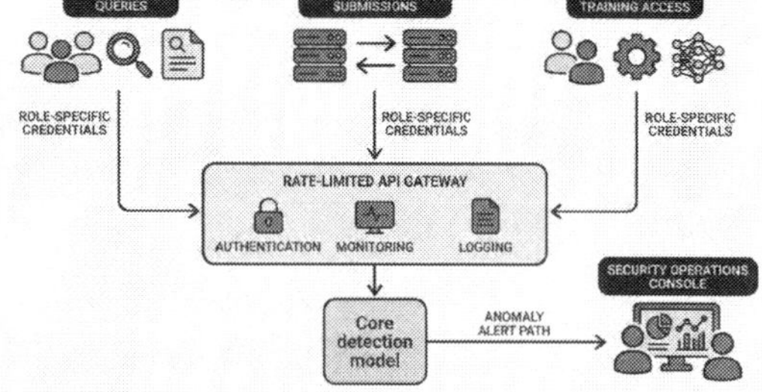

Secure architecture regulates Model API access using **role-based authentication**, rate limiting, and continuous monitoring to protect intellectual property.

The trained weights of an AI security model represent significant intellectual property and operational value for both the vendor that built the model and the adversary seeking to replicate or defeat it. Model encryption protects weights at rest and in transit, preventing an attacker who gains access to storage or network traffic from directly extracting the model architecture and parameters. Encrypted model serving, in which model weights remain encrypted during inference via hardware-level trusted execution environments, provides the strongest protection but requires infrastructure investment that most deployments have not yet made.

Model watermarking embeds identifiable signals into the model's behavior or weights that persist through copying and allow attribution if a stolen model appears in adversary infrastructure or a competing product. Watermarking does not prevent theft but enables detection and legal action after the fact. For vendor-deployed models, watermarking also provides a mechanism for detecting when a customer's access has been compromised and used to extract a copy of the model.

From a procurement standpoint, managers should ask vendors how they protect their models against extraction and which indicators they monitor for signs of systematic probing. A vendor who cannot answer these questions has not thought seriously about the security of the security tool they are selling. Contract provisions requiring vendors to notify customers when their model has been subject to significant extraction activity, or when a compromised copy is detected in the wild, represent a reasonable minimum standard for high-value AI security tool procurement.

Adversarial Robustness as a Security Design Requirement

Adversarial robustness, the ability of a model to maintain correct behavior when presented with inputs designed to cause misclassification, is not an optimization-phase concern. It is a security requirement that must be specified before a model enters development and verified before it enters production. A detection model that reliably identifies current malware variants but can be evaded by slightly modified inputs is functionally broken from a security standpoint, even if its benchmark performance is excellent.

Testing adversarial robustness requires a red team capability that most vendor evaluation processes lack. Standard penetration testing evaluates the security of the deployment infrastructure, authentication, authorization, and network controls, but not the adversarial robustness of the model itself. Organizations procuring AI security tools should require vendors to provide adversarial robustness test results from a methodology that includes adaptive attacks: attacks optimized specifically against the vendor's model, rather than generic perturbation benchmarks that any model with reasonable performance will handle correctly.

The operational implication is that adversarial robustness degrades over time as adversaries learn more about the model. A model that is robustly accurate on deployment day becomes less robust as adversaries accumulate knowledge through probing and operational observation. Robustness must be maintained through continuous adversarial evaluation and model updating, not certified once at deployment and assumed to hold indefinitely.

5.2 Training Data Integrity and Supply Chain Security

Data Provenance Tracking and Validation Pipelines

A model is only as trustworthy as the data it was trained on. Training data provenance, the documented chain of custody from the data source to the training pipeline, is the foundation for assurance that a model does what it is claimed to do. Without provenance documentation, there is no basis for confidence that training data was not tampered with, mislabeled, or contaminated before it entered the model-building process. For security-critical models, the absence of provenance documentation should be treated as a disqualifying deficiency in vendor assessment.

Validation pipelines serve as the technical counterpart to provenance documentation. Before data enters training, automated validation should check for statistical anomalies that might indicate contamination, such as unexpected label distributions, unusual feature correlations, and samples inconsistent with the claimed source characteristics. These checks catch accidental data-quality issues as well as deliberate poisoning; the same pipeline serves both data-engineering and security purposes.

The challenge of provenance tracking increases significantly when training data includes third-party contributions, threat intelligence feeds, public malware datasets, and labeling contributions from external security researchers. Each external data source introduces supply chain risk that must be managed through contractual controls, technical validation, and sampling-based human review. Organizations that rely on large public datasets without provenance assurance have introduced attack surfaces into their security models that may not become apparent until adversaries exploit them.

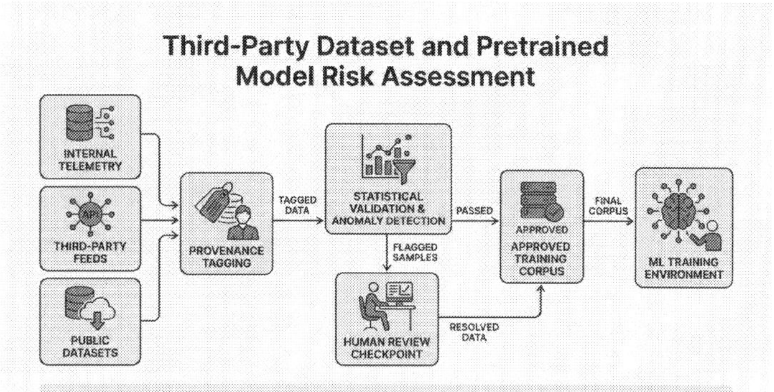

Comprehensive data lineage and validation minimize third-party risks and ensure trustworthy AI development.

Pretrained models, large foundation models fine-tuned for security applications, have become a common component of AI security tool development.

They offer significant capability advantages: the computational resources required to train from scratch are beyond the reach of most organizations, and pretrained models encode general knowledge that accelerates specialization for security tasks. They also introduce supply chain risk at the training layer, which the downstream organization that incorporates them cannot fully mitigate.

The risk profile of a pretrained model depends on the transparency of its development: what data was used, how it was curated, what fine-tuning was applied, and what evaluation was conducted before release. A pretrained model from a reputable developer with published training details, reproducible evaluation benchmarks, and an active community conducting adversarial evaluation carries substantially lower supply chain risk than an opaque model from an unverified source. The supply chain risk assessment for AI tools must include evaluation of every pretrained component incorporated in the tool's development, not only the final product.

Backdoor attacks, where a malicious actor inserts a trigger during training that causes predictable misclassification when the trigger is present, are the most severe form of contamination of pretrained models. Detection is technically challenging because backdoored models perform normally on standard evaluation benchmarks; malicious behavior manifests only when a specific trigger pattern appears in an input. Trigger-based backdoors have been demonstrated in security-relevant models, including malware classifiers and intrusion detection systems. Procurement processes that do not include backdoor-specific evaluation are not adequate for high-trust security deployments.

Continuous Monitoring for Dataset Drift and Contamination

Training data problems do not always manifest at model deployment. Dataset drift, a gradual change in the statistical properties of incoming data relative to training data, degrades model performance over time as the operational environment evolves away from the conditions under which the model was built. In security contexts, drift often reflects deliberate adversary adaptation. As attackers modify their techniques to evade detection, the distribution of malicious inputs shifts away from the training distribution, reducing detection rates on novel attack variants.

Monitoring for dataset drift requires maintaining statistical baselines for the inputs the model receives in production and comparing live distributions against training baselines at regular intervals. Significant drift, particularly toward reduced confidence in inputs that should be high-confidence, is an early indicator that the model is becoming less effective and needs retraining or augmentation. Organizations that monitor model performance only through aggregate detection metrics will miss drift signals until the degradation has become operationally significant.

Post-deployment contamination represents a distinct concern for models that incorporate feedback loops from operational data. If analyst corrections, threat intelligence updates, or detection outcomes feed back into ongoing training, adversaries who can influence these signals can gradually bias the model toward their desired behavior. Feedback pipelines must be treated as security-sensitive inputs and subject to the same validation controls as initial training data.

5.3 Inference Pipeline Protection

Input Validation and Sanitization for AI-Consuming Applications

Every AI security system has an input layer, the point at which raw data enters the system before processing by the model. This layer is the first and most accessible point for inference-time attacks. Input validation applies conventional software security principles to the specific data types that AI models consume, such as network packet captures, file samples, log events, endpoint telemetry, or natural language text, depending on the application. Validating that inputs conform to expected format, size, encoding, and source characteristics catches many attack attempts before they reach the model.

Sanitization goes beyond validation: it normalizes inputs to remove features that might carry adversarial perturbations, without changing the legitimate information content. Adversarial perturbations often consist of carefully crafted noise or minor modifications that are invisible to human observers but exploit the model's decision boundary geometry. Normalization standardizes image resolution, strips metadata, normalizes byte sequences, and applies feature-extraction pipelines that discard unusual encoding variations, reducing the attack surface for perturbation-based evasion without degrading detection capability against genuine threats.

The practical challenge is that sanitization must be calibrated to the specific attack surface. Aggressive normalization that removes all unusual features will also remove legitimate signals from novel threat variants that are, by definition, unusual. The goal is to remove features that are adversarially useful but informationally redundant, while preserving those that carry a genuine detection signal. This calibration requires ongoing adversarial evaluation to verify that the sanitization design achieves its intended effect without introducing detection gaps.

Diagram 5.4: Inference Pipeline Defense Layers

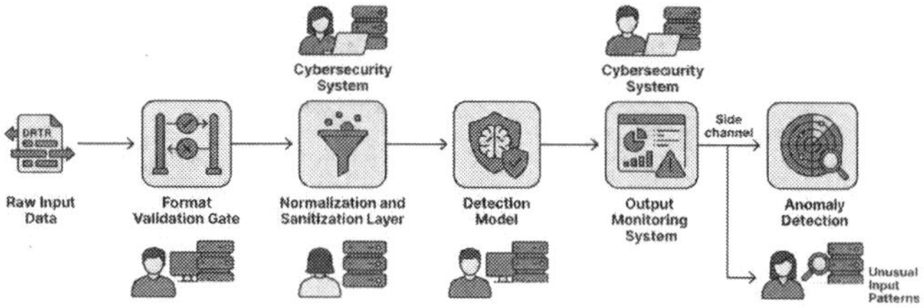

The defense-in-depth architecture of an AI inference pipeline, defense-in-depth architecture on AI inference pipeline.

AI system outputs are security observables just as network traffic and authentication events are. A detection model whose output distribution shifts unexpectedly, suddenly classifying more inputs as benign, producing unusually high or low confidence scores, or generating outputs inconsistent with historical patterns for similar inputs may be exhibiting signs of successful evasion, model degradation, or attempted manipulation. Output monitoring creates a secondary layer of anomaly detection that operates on the system's behavior rather than the inputs it processes.

Effective output monitoring requires establishing expected-behavior baselines for each model in production: typical output distributions, confidence-score ranges, classification-rate patterns, and relationships between input characteristics and output behavior. Deviations from these baselines trigger investigation rather than automatic action, because many legitimate factors, such as changes in the threat landscape, software updates, and shifts in the operational environment, can cause output changes that are not security incidents. The objective is to surface changes that warrant review, not to automate responses to every statistical deviation.

Output monitoring also provides an accountability infrastructure. When a model makes a consequential decision, blocking a file, isolating an endpoint, or flagging a user for investigation, the output record, including confidence scores and contributing features, creates an audit trail that supports both operational review and compliance requirements. Organizations that cannot explain why their AI security system made a specific decision will face increasing difficulty satisfying regulatory expectations for transparency in automated decision-making.

Infrastructure Hardening for AI Compute Environments

AI inference and training workloads run on infrastructure with distinct security characteristics. GPU clusters, high-memory compute nodes, and the storage systems holding model weights and training data represent high-value targets that must be hardened with the same rigor applied to other critical infrastructure components. The concentration of compute and data in these environments, necessary to meet the performance requirements of large-scale AI workloads, creates high-consequence breach scenarios if access controls are not proportionally strong.

Network segmentation for AI compute environments should reflect the sensitivity of the workloads they run. Training infrastructure, which holds both model weights and training data, warrants the most restrictive segmentation, ideally isolated from internet-accessible systems with access mediated through bastion hosts and limited to specific authenticated users performing specific tasks. Inference infrastructure handling production security workloads requires different trade-offs: lower latency demands more accessible network paths, but this accessibility must be offset by compensating controls, including comprehensive traffic logging and behavioral monitoring.

Supply chain hardening extends to the software stack running on AI compute infrastructure. ML frameworks, GPU drivers, container base images, and model serving platforms all represent attack surfaces for dependencies. Organizations that do not maintain a software bill of materials for their AI compute stack and monitor components for known vulnerabilities have introduced supply chain risk at the infrastructure layer, complementing the model-level risks described earlier in this chapter.

Diagram 5.5: AI Compute Infrastructure Security Zones

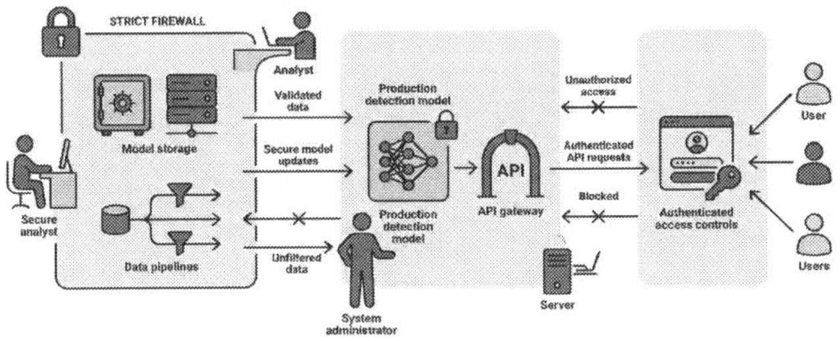

Robust network segmentation isolates critical AI training and model assets from potential external threats.

Security Requirements in AI System Procurement and Contracting

Procurement of AI security tools requires a different evaluation framework than procurement of conventional security software. The traditional checklist feature set, integration compatibility, vendor reputation, and pricing must be extended to include AI-specific security properties: adversarial robustness, training data provenance, extraction resistance, output monitoring capability, and the vendor's security incident response process for AI-specific attacks. Without explicit evaluation of these properties, procurement decisions will systematically favor tools that perform well on benchmarks but may be highly vulnerable to sophisticated adversarial attack.

Contract terms for AI security tool procurement should specify minimum security standards for the tool itself, not only for the vendor's general security practices. Relevant provisions include notification requirements when the vendor's model has been subject to extraction attacks or when evasion toolkits targeting the model are detected in the wild; access controls and audit rights for the vendor's model development and training infrastructure; and requirements for ongoing adversarial testing with results made available under appropriate confidentiality terms. These provisions are unusual in current procurement practice but represent the direction in which responsible AI tool procurement is moving.

Organizations with significant AI security investments should consider requiring vendors to provide an AI system security disclosure, a structured summary of model development practices, training data sources, adversarial evaluation results, and known limitations. This disclosure, analogous to a software security vulnerability disclosure but specific to AI system properties, enables informed procurement decisions and holds vendors accountable for claims about the capabilities and resilience of their tools.

Red Team and Adversarial Testing Programs for AI Systems

Red team exercises for AI security systems require capabilities that extend beyond traditional penetration testing. A red team evaluating adversarial robustness must be able to construct adaptive attacks designed specifically to evade the model under test, rather than applying generic benchmark attacks that may not reflect the adversary's actual capabilities against that model. This requires team members with machine learning expertise who can reason about model decision boundaries, construct meaningful perturbations, and evaluate the practical operational value of discovered evasion techniques.

Red team programs should include both white-box evaluation, where the team has full access to the model architecture and weights, and gray-box evaluation, which simulates the access an external adversary with API access would have. White-box evaluation provides the most technically complete picture of model vulnerabilities; gray-box evaluation more accurately represents real-world attack conditions. Both are necessary for a complete adversarial assessment.

The output of AI red team exercises must include not only discovered vulnerabilities but also practical evasion techniques and the estimated operational cost to an adversary of exploiting each vulnerability. A theoretical evasion that requires months of computation and thousands of API queries is less operationally significant than one achievable in hours with commodity resources. Prioritizing remediation based on operational exploitability, rather than theoretical severity alone, produces more effective security investment.

Diagram 5.6: AI Red Team Program Structure

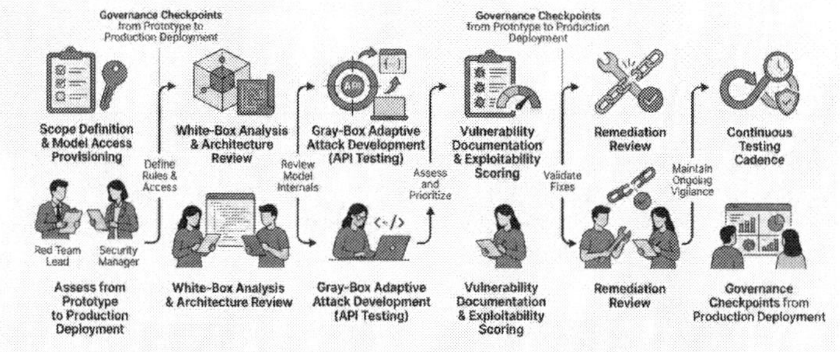

Structured AI red teaming validates security from prototype to production through rigorous analysis, targeted attacks, and validated remediation.

AI security systems go through a development lifecycle with distinct security-relevant phases: data collection and curation, model development and training, internal evaluation, pilot deployment, and full production release. Each phase introduces different security risks and requires different governance controls. Organizations that apply security review only at the final deployment gate miss the opportunity to catch training-time vulnerabilities such as supply chain contamination, backdoor implantation, and inadequate adversarial robustness before they are baked into a production system.

A governance checkpoint in the data curation phase evaluates provenance documentation, validation pipeline outputs, and the results of any statistical anomaly-detection run against the training corpus. This checkpoint creates a formal approval record that establishes the basis for downstream assurance claims. If the training data is later found to be contaminated, the checkpoint record establishes whether the contamination entered before or after the approved phase, providing a basis for incident scope assessment.

Production deployment checkpoints should include adversarial robustness evaluation against the most current threat techniques available at the time of deployment, infrastructure security review of the compute environment, and formal sign-off from both security and data science leadership. The dual sign-off requirement acknowledges that production AI systems sit at the intersection of two disciplines, whose quality standards must be met before systems enter service. Post-deployment governance includes scheduled robustness re-evaluation, output monitoring review, and a defined process for triggering retraining when drift or degradation indicators reach defined thresholds.

5.4 Manager's Checklist

- Require vendors of AI security tools to provide documentation of training data provenance, adversarial robustness evaluation results, and model extraction protection mechanisms before procurement approval.

- Establish role-based API access controls for every AI model in production, ensuring that query permissions are scoped to the minimum necessary for each authorized use case and that all queries are logged with sufficient detail to detect extraction patterns.
- Implement rate limiting and query pattern monitoring on all model APIs, with alert thresholds calibrated to distinguish legitimate operational query volumes from systematic probing activity.
- Include pretrained model and third-party dataset risk assessment in your AI procurement checklist, requiring vendors to document all pretrained components and the provenance of third-party training data.
- Establish output monitoring baselines for AI security systems in production and define deviation thresholds that trigger review, ensuring that shifts in model output distributions surface for investigation rather than going undetected.
- Segment AI compute infrastructure from general enterprise networks, applying access controls appropriate to the sensitivity of model weights and training data, and maintaining a software bill of

materials for all AI compute stack components.

- Build AI-specific red team capability either internally or through specialized vendors and conduct adversarial robustness exercises that include adaptive attacks against your specific deployed models at least annually.

8. Define governance checkpoints across the AI development lifecycle with formal approval requirements at data curation, training completion, and production deployment stages, documented in a way that supports audit and incident response.

5.5 Security Cannot Exempt Itself

The irony embedded in this chapter's subject matter should not obscure its seriousness. The systems deployed to defend the enterprise are high-value targets with novel attack surfaces, and the organizations that deploy them have largely not invested in defending them against the attacks they face. The adversary who successfully maps the decision boundaries of a detection model, poisons the training data of a behavioral analytics system, or compromises the compute infrastructure running security workloads, gains leverage not despite your security investment but because of it.

The practices described here, access controls for model APIs, training data provenance tracking, adversarial robustness evaluation, inference pipeline hardening, and supply chain security for pretrained components, are not exotic security research. They are the application of established security engineering principles to a new system category. Organizations that consistently apply these principles will build an AI security infrastructure that is genuinely more difficult to subvert. Those that do not will continue deploying tools whose effectiveness degrades as adversaries learn to exploit their specific weaknesses.

The governance layer is ultimately what makes AI security sustainable. Technical controls without governance decay; governance without technical controls is theater. Building an AI-secure development lifecycle with defined checkpoints, documented assurance chains, and clear accountability for both data science and security quality creates the institutional discipline needed to maintain a defensible AI security infrastructure over time. In a field where both the threats and the defensive tools are evolving continuously, that discipline is the most durable investment an organization can make.

6 The Transformed Operations Center

The security operations center of a major defense contractor ran three shifts around the clock. On any given shift, four analysts monitored a console that displayed between eight hundred and twelve hundred active alerts, the output of a detection stack that had grown steadily more capable and correspondingly more voluminous over five years. During busy periods, the queue grew faster than analysts could clear it. The practical consequence was triage by exhaustion: after the first two hours of each shift, the probability that a given alert received genuine analytical attention dropped sharply. Analysts developed pattern-matching shortcuts to clear volume. New alerts were judged primarily by resemblance to recently resolved items, not by independent evaluation.

When the organization deployed an AI triage platform, the change was not primarily technological. The platform absorbed approximately 90% of alert volume through automated enrichment, correlation, and classification, returning to analysts a queue of 20 to 40 genuinely ambiguous cases per shift rather than hundreds of mechanically patterned ones. Analysts who had spent most of their time confirming that routine alerts were indeed routine began spending most of their time investigating the cases the system had correctly identified as deserving human attention. Detection quality improved. Response times for genuine incidents fell. Analyst retention, which had been eroding, stabilized.

The lesson is not that AI replaces analysts. It is that a significant fraction of what analysts had been doing was not genuinely analytical work, and that restoring their capacity for genuine analysis by removing the routine mechanical work had measurable operational consequences.

Diagram 6.1 Alert Volume Before and After AI Triage

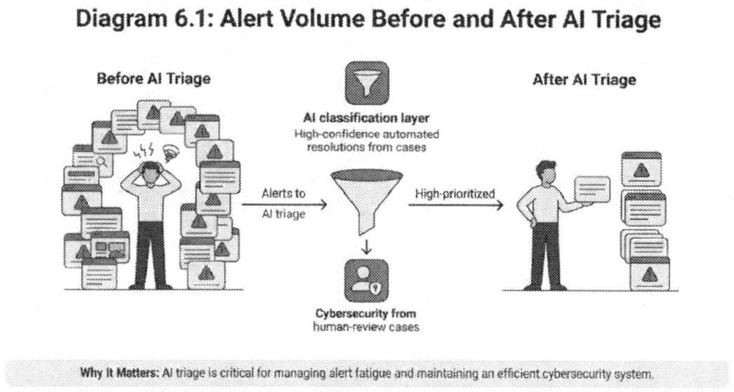

Alert fatigue is not a morale problem. It is an operational failure mode with direct consequences for the organization's ability to detect and respond to genuine threats. When analysts process volume rather than analyze content, the cognitive conditions required for genuine threat detection, careful attention, contextual reasoning, and pattern recognition across disparate signals are systematically absent. The investment in detection technology that generates the alerts becomes partially wasted because the human stage of the detection pipeline cannot process what the technology produces.

The problem compounds over time. Analysts who spend prolonged periods in high-volume, low-signal environments develop coping mechanisms that persist even when the environment improves. Abbreviated analysis, over-reliance on automated recommendations, and reduced engagement with ambiguous cases are adaptations to volume pressure that do not automatically reverse when volume decreases. The human capital consequences in reduced analytical quality, increased attrition among experienced staff, and difficulty attracting capable new analysts to a role perceived as mechanical persist beyond the immediate operational impact.

For managers, the AI-enabled SOC transformation is an opportunity that requires deliberate design to realize. Technology can change the quantity and character of work reaching analysts. Still, it will not automatically change how analysts approach their work, how teams are organized, how performance is measured, or how roles are defined. Organizations that deploy AI triage and then leave everything else unchanged will capture only a fraction of the available benefit. Realizing the full value of AI-augmented operations requires rethinking the operations center as an institution, not only its tooling.

6.1 Alert Fatigue and the Crisis of Attention

Quantifying the Operational Cost of Alert Overload

Alert overload has specific, measurable operational costs that managers can quantify to build the business case for AI triage investment and to measure the benefit after deployment. The most direct cost is analyst time: if an analyst spends 40 seconds per alert and receives 400 alerts per shift, three-quarters of the shift has been consumed by alert processing before any investigative work begins. At that ratio, the cost of alert volume is not merely opportunity cost; it actively displaces the activity the SOC exists to perform.

A second quantifiable cost is the mean time to detect genuine incidents. When analysts are working through a backlog rather than investigating current alerts, incidents that arrive during high-volume periods wait behind a queue of low-priority items for human review. Measuring the distribution of alert queue depth at the time genuine incidents enter the system, rather than mean queue depth across all periods, often reveals that incidents disproportionately wait in queues precisely because adversaries time their activities to coincide with high-volume periods, exploiting the known limitation of volume-overwhelmed defenders.

A third cost, less often quantified but organizationally significant, is the false-negative rate as a function of analyst workload. Research across high-stakes monitoring domains consistently finds that error rates increase with sustained high-throughput work, particularly for detecting anomalies that require comparison to context rather than simple pattern matching. Documenting the relationship between queue depth and false-negative rate in your own environment through retrospective analysis of incidents detected late or missed entirely provides a direct cost measure of the operational impact of alert overload.

How Noise Degrades Analyst Performance and Retention

The analytical degradation caused by sustained high-alert-volume environments follows a recognizable pattern. Early in a high-volume shift, analysts engage with alerts as distinct items. As volume pressure increases, attention shifts from content evaluation to category recognition, allocating time based on alert type and source rather than on careful reading of each item. By the latter portion of high-volume shifts, the primary cognitive activity is category sorting rather than threat analysis. This is not a failure of individual analysts; it is a predictable response to an impossible task volume that any competent professional would exhibit under similar conditions.

Retention consequences accumulate gradually but follow a consistent pattern. The analysts who leave first are typically the most experienced and capable, those who have external market value to find positions where their skills are more fully engaged. The analysts who remain are disproportionately those with fewer alternatives, creating a selection effect that degrades analytical quality independently of the volume problem. By the time attrition becomes visible as a staffing crisis, the quality degradation has often been underway for months or years.

Manager visibility into analyst cognitive states is limited in most SOC environments. Quantitative performance metrics, such as alerts processed, tickets closed, and mean time to respond, are straightforward to collect but measure throughput rather than quality. An analyst processing three hundred alerts per shift may appear highly productive by throughput metrics while missing a sophisticated attack hidden among routine items. Building quality metrics into performance assessment, sampling analyst reasoning on a portion of cases, and conducting retrospective analysis of incident handling provides the visibility needed to detect analytical degradation before it leads to security failures.

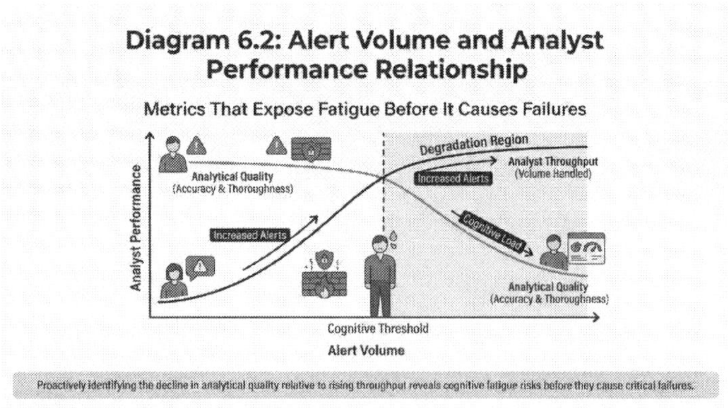

Diagram 6.2: Alert Volume and Analyst Performance Relationship

Proactively identifying the decline in analytical quality relative to rising throughput reveals cognitive fatigue risks before they cause critical failures.

Proactive fatigue detection requires metrics that measure analytical quality and engagement rather than throughput. Several indicators reliably precede the operational failures that alert fatigue eventually produces: declining variance in analyst decisions as cases trend toward uniform disposition regardless of content; increasing alignment between analyst decisions and default automated recommendations without documented override reasoning; shorter dwell times on individual alerts across the distribution, including the complex cases that should require longer consideration; and increasing use of boilerplate documentation language that signals template application rather than genuine case analysis.

Shift-level analysis of these indicators, comparing performance across shift segments and alert-volume conditions, provides a leading indicator of analytical degradation. A shift where analyst dwell times drop uniformly across all alert categories, or where override rates on automated recommendations fall to near zero, is a shift where the cognitive conditions for genuine analysis are no longer present. Identifying these patterns before they persist and before they produce missed incidents allows managers to intervene through staffing adjustments, queue management, or analyst rotation.

The investment required for these metrics is not primarily technical; most SOAR platforms can generate the underlying data if configured to capture decision timestamps, documentation length, and override events. The investment is primarily managerial: deciding that quality metrics matter, designating someone to monitor them, and building management practices that respond to quality signals rather than prioritizing throughput over all other considerations.

6.2 AI-Powered Triage and Case Management

Automated Alert Enrichment and Contextual Correlation

AI triage systems deliver their first value through enrichment and correlation, transforming raw alerts from isolated signals into contextualized cases before human review begins. Enrichment automatically attaches relevant context to each alert: the historical behavior profile of the associated user or entity, the asset inventory record for the affected system, the threat-intelligence reputation of observed indicators, and the prior incident history of related accounts and systems. Analysts who spend significant time manually gathering this context receive it, assemble and format it, as the case enters their queue.

Correlation goes beyond enrichment to identify relationships among alerts spanning time, entities, and detection systems. A sophisticated attack rarely generates a single alert; it produces a sequence of low-confidence signals across multiple monitoring systems that individually appear routine but collectively indicate malicious activity. Correlation engines, whether rule-based or ML-driven, identify these patterns and surface them as single cases rather than as separate alerts, reducing the cognitive burden of cross-signal analysis that is particularly susceptible to failure when analysts are processing high volumes.

The quality of enrichment depends on the quality of the underlying data sources: asset inventories must be current, user behavior baselines must be maintained, and threat intelligence feeds must be actively managed rather than passively subscribed. Organizations that deploy AI triage without investing in the data foundations it depends on will find that enriched cases are no more useful than raw alerts. The system will accurately report that it cannot find a nonexistent context. Enrichment quality improvement is therefore often a data infrastructure project as much as an AI deployment project.

Diagram 6.3: Alert Enrichment and Correlation Workflow

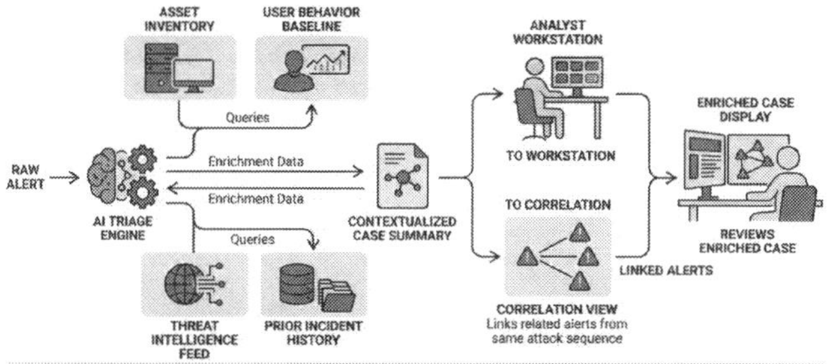

Automated enrichment and correlation streamline investigation, empowering analysts with comprehensive context.

Risk-Prioritized Case Queues and Analyst Workflow Design

AI triage is most effective when the case queue it produces is designed to match analyst cognitive capacity rather than ranked by algorithmic confidence scores. A queue that presents the highest-confidence automated resolutions alongside the most ambiguous human-review cases in an undifferentiated stream has not improved the analyst's working environment; it has substituted one form of sorting work for another. Effective queue design presents analysts with a workload structure that matches their attention capacity and task-switching costs.

The practical design question is how to segment case queues to support analyst workflow. One approach distinguishes between high-confidence positive cases where automated detection has strong support and human review confirms or closes them and genuinely ambiguous cases that require investigation. Batching high-confidence cases into a rapid-review workflow, where analysts confirm automated assessments, leverages the speed with which trained analysts process well-supported findings. Routing genuinely complex cases to a separate investigative workflow preserves analyst cognitive capacity for the work that most requires it.

Analyst workflow design should be tested against the actual distribution of cases in the production environment rather than designed from theoretical principles. Case complexity distributions in SOC environments are often more bimodal than they appear in aggregate metrics. In most cases, the majority are routine, a small minority are genuinely complex, and the distinction between these categories is often recognizable to experienced analysts but not well captured by simple confidence scores. Workflow design that maps onto the actual case distribution will consistently outperform a design based on assumptions about that distribution.

Continuous Learning: How Triage Models Improve with Analyst Feedback

AI triage systems that incorporate analyst feedback into ongoing model improvement create a learning dynamic that conventional rule-based systems cannot replicate. When analysts confirm, override, or escalate automated recommendations, their decisions convey information about the model's accuracy on specific case types and about the features that distinguish cases the model handles well from those it gets wrong. Capturing this feedback systematically and feeding it into model retraining produces a system that improves continuously with operational use rather than degrading as the threat environment evolves.

The feedback loop design requires attention to several failure modes. Selection bias arises when analysts disproportionately provide feedback on cases that they find interesting or unusual, producing a feedback distribution that overrepresents rare cases and underrepresents the routine majority. This can cause the model to improve on edge cases while degrading on its core workload. Capturing feedback on a representative sample of cases, including routine confirmations, overrides, and escalations, maintains a feedback distribution that reflects the actual operational environment.

Analyst gaming of automated recommendations represents a second failure mode. If analysts learn that consistently disagreeing with automated recommendations, or consistently confirming them, produces workload outcomes they prefer fewer complex cases, faster queue clearance, they will adjust their behavior accordingly, contaminating the feedback signal. Incentive structures around analyst feedback must be designed to reward accurate assessment rather than throughput optimization, and the feedback data should be monitored for patterns that suggest gaming rather than genuine evaluation.

DIAGRAM 6.4: **TRIAGE MODEL FEEDBACK AND CONTINUOUS LEARNING LOOP**

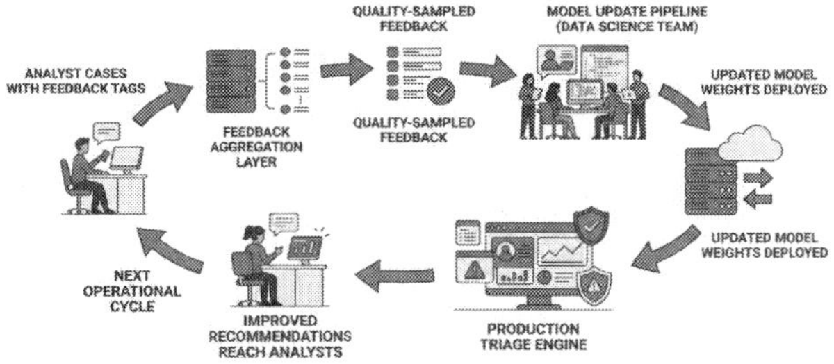

Continuous feedback and model updates refine cybersecurity triage accuracy over time.

Playbook Architecture for High-Confidence Automated Actions

Automated response capability is the logical extension of automated triage. If a system can reliably classify a case as high-confidence malicious, it can also initiate proportionate response actions without waiting for human authorization. The operational benefit is response speed. Automated containment can occur in seconds rather than the minutes or hours required for human review, and the capacity of automated response scales with detection volume rather than with analyst headcount. The risk is automated error, where a high-confidence but incorrect classification triggers containment actions against legitimate activity.

Playbook design for automated response begins with classifying potential actions by reversibility and impact. Low-impact, fully reversible actions, such as quarantining a suspicious file, temporarily blocking a suspicious network connection, and sending an automated user notification, can be authorized for fully automated execution at relatively low confidence thresholds because the cost of a false positive is a minor operational inconvenience. High-impact or difficult-to-reverse actions such as isolating an endpoint from the network, locking a user account, or terminating a system process require higher confidence thresholds and typically benefit from a human confirmation step before execution.

Playbook architecture should be treated as a living document, updated as experience accumulates about the reliability of automated decisions in specific case categories. Actions initially requiring human confirmation may be moved to full automation as track records establish confidence in classification accuracy. Actions initially implemented with fully automated execution may be moved to a confirmation-required status if the false-positive experience exceeds acceptable thresholds. This iterative approach builds trust in automated responses through operational evidence rather than presuming confidence in advance.

Balancing Speed of Response Against Risk of Automated Error

The speed-error trade-off in automated response is a management decision, not a technical one. The technology can be configured to act faster with more automation or more carefully with more human checkpoints; the choice of where to set those controls reflects the organization's assessment of the relative consequences of a delayed response to a genuine threat versus an erroneous automated action against legitimate activity. This assessment differs across organizations. A financial institution processing millions of transactions will have a different tolerance profile than a manufacturing facility with a small number of critical operational systems. It must be revisited as the operational environment changes.

A useful framework for this trade-off is to quantify the operational cost of each failure mode separately and compare them directly. The cost of a delayed response is measured by the additional damage an adversary can cause per unit of response-time delay, endpoint-compromise depth, lateral-movement extent, and data-exfiltration volume. The cost of an automated error is measured by the disruption to legitimate operations caused by an incorrect containment action, user productivity loss, business process interruption, and incident response overhead to investigate and reverse the action. Comparing these costs directly, weighted by the probability of each failure mode, produces a quantified basis for confidence threshold decisions rather than a qualitative judgment about risk tolerance.

Monitoring the operational consequences of automated actions in both directions, tracking cases where automated response successfully contained a genuine threat faster than human response could have, and cases where automated response produced false-positive disruption, provides the operational data needed to recalibrate thresholds over time. Organizations that implement automated responses without this monitoring cannot know whether their threshold settings are achieving their intended effect or whether they are systematically over-automating or under-automating relative to their actual risk environment.

Diagram 6.5: Automated Response Confidence Threshold Framework

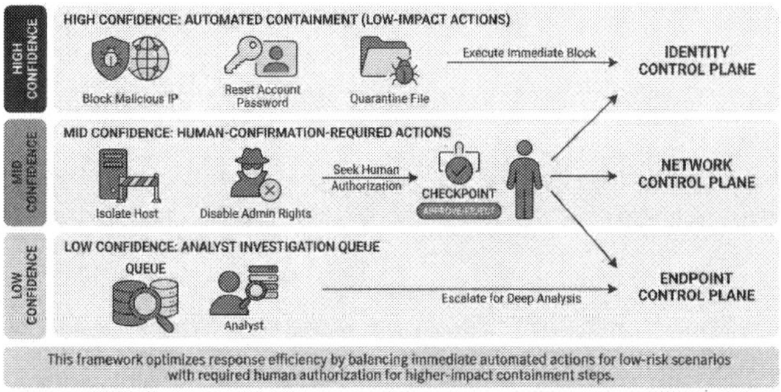

Automated response capability depends on integration with the control planes that can execute containment actions: the identity management system that controls authentication and access, the network control systems that can block or redirect traffic, and the endpoint management systems that can isolate or remediate devices. Each of these integrations must be engineered with the integration's security in mind. A SOAR platform with broad endpoint control capabilities is a high-value target whose compromise would enable an attacker to execute automated containment actions at scale.

Integration architecture should follow least-privilege principles: the SOAR platform should have only the control capabilities required for its defined playbook actions, not broad administrative access to the control planes it integrates with. Each automated action should be executed through a dedicated, narrowly scoped service account rather than through administrative credentials that would confer broader access if compromised. These constraints require more careful integration engineering than granting the SOAR platform administrative access to all connected systems. Still, they substantially reduce the blast radius if the automation platform is itself compromised.

Testing integrations in a representative lab environment before production deployment, and maintaining a test environment for ongoing validation afterward, are essential for the reliability of automated responses. An automated playbook that fails silently when an integration is degraded, producing no response to a genuine threat, is worse than no automation because it creates false confidence that a response has occurred. Integration health monitoring, with alerting when connected control planes become unavailable or degraded, is a prerequisite for operational automation that the SOC team can rely on.

6.3 SOC Organizational Design for the AI Era

Role Evolution: From Tier-1 Analyst to Threat Hunter

The traditional SOC tier structure, Tier 1 handling alert triage, Tier 2 investigating escalated cases, Tier 3 performing advanced threat hunting and engineering, was designed around the distribution of volume and complexity in a pre-AI alert environment. AI triage substantially reduces the volume of work at Tier 1, and the nature of the remaining work shifts from mechanical sorting to contextual evaluation. Simultaneously, AI-assisted threat-hunting tools reduce the technical barrier to sophisticated hunting, making threat hunting accessible to analysts who previously lacked the tooling and data access required for Tier 3 activities.

The organizational implication is a compression of the tier structure. When AI handles the volume previously defined as Tier 1, and when hunt tooling extends meaningful threat-hunting capability to a broader analyst population, the distinction between Tier 1 and Tier 2 becomes less operationally meaningful. More organizations are moving toward a flatter role structure, with an emphasis on analytical depth, such as case investigator roles that combine alert evaluation, incident investigation, and threat hunting within a single job function, rather than a tiered model that segregates these activities by seniority.

This role evolution has significant workforce implications. Analysts who entered the field through Tier 1 roles, which have historically served as the primary entry point into security operations, face a different career path in an AI-augmented environment. Organizations that eliminate Tier 1 positions without redesigning the entry-level career pathway will find themselves unable to develop the analyst pipeline that senior roles depend on. Designing AI-era entry roles that provide genuine analytical challenges and clear development pathways is a talent strategy requirement, not just an HR consideration.

Diagram 6.6: SOC Role Evolution: Tiered to Integrated Staffing Models and Coverage Strategies with AI Augmentation

AI augmentation changes the staffing math of continuous coverage operations. When AI triage handles the majority of alert volume continuously, the human staffing requirement for overnight and weekend coverage can be recalibrated, not eliminated, but optimized around the genuine human work that remains rather than the volume of alerts that would previously have required continuous manual attention. Organizations that maintain pre-AI staffing levels for off-hours coverage are overstaffing for the alert-triage work that AI now handles and may be underinvesting in day-side analytical capacity, where complex investigations and threat-hunting activities benefit most from senior staff attention.

Coverage strategy recalibration requires an honest assessment of when genuinely complex work occurs in your specific environment. Most organizations find that genuine incidents requiring expert-level analytical attention cluster during business hours, when adversaries are operating against business-hours authentication environments and when the organizational activity that generates meaningful context for incident analysis is underway. Off-hours operations benefit from automated response capabilities and reliable escalation paths to on-call senior staff, rather than from continuous staffing with junior analysts whose effectiveness at complex analysis is limited.

Hybrid coverage models, AI-continuous, human on-call with defined escalation thresholds for off-hours, full staffing during day operations, represent the emerging operational pattern for AI-augmented SOCs. Implementing this model requires defining the automated response envelope that operates without human oversight during reduced-coverage periods and establishing escalation thresholds that reliably surface genuine high-priority incidents for human review rather than accumulating them for morning triage. The reliability of these thresholds must be validated through operational testing, not assumed from design intent.

Performance Measurement in AI-Augmented Operations

Performance measurement in AI-augmented operations requires rethinking what success looks like when the volume metric that previously defined SOC productivity alerts processed per analyst per hour is no longer the binding constraint. Organizations that continue to measure analyst performance primarily through throughput metrics after deploying AI triage are measuring the wrong thing and will reward the wrong behaviors: analysts will optimize for case-closure speed rather than analytical quality, reproducing alert-fatigue dynamics even in a reduced-volume environment.

Quality-focused metrics appropriate to AI-augmented operations include: mean time to identify and scope genuine incidents, the proportion of escalated cases that result in confirmed incidents versus false escalations, override accuracy measured against subsequent investigation outcomes, and the frequency with which analyst investigations discover context that updates the AI model's assessment rather than simply confirming it. These metrics measure the analytical contribution analysts make to security outcomes rather than the volume of items they process.

Team-level performance measurement should include the contribution of the human-AI system as a whole, not only individual analyst productivity. Relevant team metrics include the rate at which genuine incidents are detected within defined time windows, the false positive rate of automated response actions, and the quality of case documentation that feeds the continuous learning loop. Measuring these outcomes rather than activity inputs creates accountability for the results the AI-augmented team produces together, rather than for the volume of work each processes in isolation.

6.4 Manager's Checklist

- Measure the current operational cost of alert overload in your SOC: calculate the percentage of analyst time consumed by alert processing versus investigation, estimate queue depth at the time of confirmed incidents, and assess analyst attrition rates against industry benchmarks.
- Evaluate AI triage platforms on enrichment quality and correlation capability, not only on alert volume reduction claims; verify that the underlying data foundations (asset inventory, user behavior baselines, threat

intelligence) are current enough to support meaningful enrichment.

- Design analyst case queues and workflow structures that match cognitive capacity to case complexity, distinguishing rapid-review workflows for high-confidence automated resolutions from investigative workflows for genuinely ambiguous cases.
- Build a feedback capture mechanism into your SOAR or triage platform that records analyst decisions with sufficient context to support model improvement, and designate ownership of feedback quality monitoring.
- Develop a playbook architecture for automated response that classifies potential actions by reversibility and impact, establishes confidence thresholds for each action category, and documents the rationale for threshold choices as a baseline for subsequent review.
- Harden integrations between your automation platform and control planes (identity, network, endpoint) using least-privilege service accounts scoped to playbook-required actions and implement integration health monitoring with alerting on degraded connectivity.
- Assess whether your SOC role structure and career pathways are designed for the AI-augmented operating model or the pre-AI volume model, and identify whether

entry-level analyst roles provide the analytical development and career progression needed to build your senior analyst pipeline.

- Redefine SOC performance metrics to emphasize analytical quality and outcome achievement rather than throughput, and review whether current incentive structures reward the analytical behaviors that AI-augmented operations require.

Diagram 6.7: AI-Augmented SOC Operational Model Overview

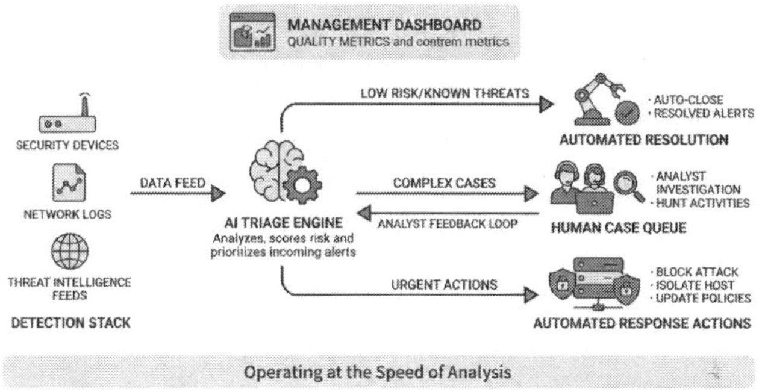

The operations center transformation described in this chapter is not primarily about reducing costs or eliminating positions. It is about making security operations actually work, closing the gap between the capability of detection technology and the capacity of human analysts to convert detection signals into security outcomes. That gap has been widening for years as detection stack sophistication outpaced the ability of human teams to process what the technology produces. AI triage is the mechanism for closing it.

Realizing that benefit requires organizational investment that matches the technological investment. Deploying AI triage without redesigning analyst workflows, case queue structures, and performance metrics will produce disappointing results, not because the technology fails, but because the human systems around it remain optimized for a volume-management model that no longer applies. The organizations that treat SOC transformation as a combined technology and organizational design challenge will see materially better outcomes than those that treat it as a software deployment.

The competitive advantage of an AI-augmented SOC is not the automation itself, as technology is available to any organization that can procure it. The advantage is the analytical culture built when experienced professionals are freed from mechanical volume work and given genuine analytical challenges. Organizations that invest in building that culture through role design, measurement, development programs, and management practices that reward analytical quality will develop institutional capabilities that persist and improve over time. Technology can be purchased; culture must be built.

7 Governing Intelligence

The compliance audit came without warning. A federal agency's inspector general, acting on new regulatory guidance around automated decision-making in government systems, requested documentation of every AI-powered tool the agency's cybersecurity division used, how each model made decisions, what data it processed, who approved its deployment, and how the organization would explain an adverse outcome to an affected employee or contractor. The CISO had deployed six AI-based detection and response tools over the previous two years, all from reputable vendors, all with strong performance metrics. What she did not have was a single page of documentation that would satisfy the inspector general's questions. The tools had been procured based on efficacy rather than accountability. Within a week, the agency placed three of the six tools on administrative hold pending review, disrupting triage workflows that analysts had spent months calibrating. The event did not result from any failure of the technology. It resulted from the absence of governance.

7.1 Why Governance Cannot Be an Afterthought

AI-powered security tools make consequential decisions at machine speed: blocking network connections, quarantining endpoints, restricting user accounts, escalating or suppressing alerts. When those decisions are wrong, when a legitimate employee is locked out of a system during a critical production window, or when an alert suppression causes an incident to go undetected, someone must be accountable. In organizations that have deployed AI without governance structures, accountability is genuinely unclear. The vendor claims the model performed as documented. The security team points to the configuration. The business unit absorbs the loss without a satisfactory explanation. This diffusion of accountability is not a coincidence. It is the predictable outcome of deploying autonomous decision-making technology without the governance architecture needed to assign responsibility clearly.

The risk is not hypothetical. Regulators in the United States, European Union, and United Kingdom have begun requiring organizations in high-stakes sectors to demonstrate that automated decision systems are explainable, auditable, and subject to human oversight. Financial services firms must now address AI model risk in examinations. Healthcare organizations deploying AI are subject to scrutiny under both privacy and safety frameworks. Federal contractors face emerging requirements from the Office of Management and Budget. Sector-agnostic requirements are arriving through employment, anti-discrimination, and consumer protection law. The organizations that treat governance as a compliance checkbox, rather than an operational discipline, will find themselves repeatedly unprepared as the regulatory environment continues to harden.

Governance matters for a reason that precedes regulation: trust. The analysts who must act on AI-generated risk scores, the executives who must allocate budget based on AI-assessed posture metrics, and the employees whose behavior AI systems monitor all need to understand what the systems do, how they do it, and what limits are placed on their authority. Organizations that cannot answer those questions have not deployed AI security programs. They have deployed black boxes, and accountability lives nowhere. Trust earned through demonstrated governance discipline is also a competitive and reputational asset. Organizations that can show regulators, partners, and customers a coherent governance framework for their AI security tools occupy a stronger position than those that cannot, independent of any specific regulatory requirement.

Diagram 7.1: The Governance Gap: From Deployment to Accountability

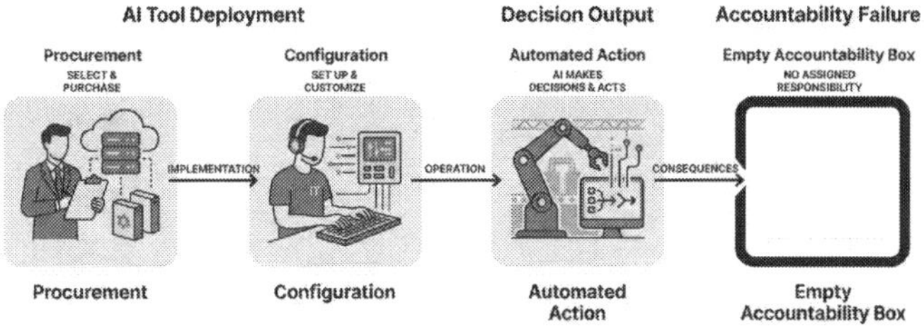

The governance gap emerges when AI systems make independent decisions without clear attribution of responsibility for their outcomes.

7.2 AI Ethics in Security Applications

Explainability Requirements for Automated Security Decisions

Explainability does not mean that every model must produce a natural-language rationale for every output. It means that a trained analyst, a compliance officer, or a regulator can understand, at an appropriate level of detail, why a system produced a particular outcome. For a behavioral analytics platform flagging a user as a high-risk insider threat, explainability means the analyst can see which behaviors contributed to the score, what baseline those behaviors were measured against, and how confident the model is in its assessment. For an automated containment action, it means there is a documented logic chain between the triggering condition and the response. Explainability at the individual-outcome level is different from explainability at the system level; organizations need both.

Managers evaluating AI security tools should require vendors to specify what form of explainability their products support. Attention-based explanations highlight which input features drove an output. SHAP values quantify each feature's contribution to a prediction. Rule extraction surfaces the implicit logic of a model in a human-readable form. Some systems provide none of these; they deliver a risk score and a recommendation without a rationale. Those systems create accountability gaps that no organizational policy can fully close. Procurement checklists must include explainability as a non-negotiable requirement, not an optional premium feature. When a vendor cannot describe the mechanism by which their system explains its outputs, the organization should treat that gap as a material procurement risk.

Internal documentation must complement vendor capabilities. When a tool flags a user, blocks a connection, or suppresses an alert, the security program should be able to reconstruct the chain of events: what the system observed, what it decided, what action followed, and who was notified. This audit trail is not a bureaucratic nicety. It is the evidentiary foundation for defending decisions to regulators, legal counsel, and employees who may challenge outcomes that affect them. Organizations that do not build this documentation practice into their AI deployments will improvise it under pressure, at the worst possible moment. The time investment to build good documentation practices during normal operations is far smaller than the cost of reconstructing decision histories from fragmented logs after an adverse event.

Bias and Fairness in Detection and Response Systems

Bias in AI security systems is a concrete operational problem, not a philosophical one. Behavioral analytics models trained on data from a predominantly office-based workforce will systematically score remote workers as higher risk, not because remote workers behave suspiciously, but because their behavior diverges from the training baseline. Models trained during periods of stable operations will misclassify users during legitimate business events such as mergers, system migrations, policy changes, or organizational restructuring. Detection systems trained primarily on one class of threat-actor behavior may perform significantly worse against other classes. Each of these biased sources causes operational degradation and potential legal exposure, without any obvious signal that something has gone wrong.

Fairness audits for AI security systems require the same rigor as performance audits. Organizations should measure false positive and false negative rates disaggregated across meaningful population segments: job function, location, employment classification, time zone, and any other dimension where systematic differences in behavior might exist. A system that performs well on aggregate metrics but consistently over-flags one class of users creates both legal exposure and operational damage. Analysts who learn that certain groups are routinely flagged for normal activity begin ignoring those alerts, degrading the system's effective sensitivity precisely where the organization thought it was protected. Correcting bias after it has shaped analysts' behavior requires both technical remediation and repair of organizational trust.

Accountability Structures When AI Systems Cause Harm

When an AI system causes harm, a false positive that locks a surgeon out of a clinical application during an emergency, or a suppressed alert that allows ransomware to propagate for twelve additional hours, the organization needs a clear answer to who is responsible. That answer must exist before the harm occurs. Accountability structures for AI security systems must assign named roles at three levels: operational accountability for the system's day-to-day performance and configuration; governance accountability for the policies and boundaries within which the system operates; and escalation accountability for decisions that exceed the system's authorized scope. Diffuse accountability produces no accountability at all. When accountability is diffuse, the natural human response is to point to someone else, and the organizational response is to commission a review that produces recommendations that nobody owns.

Accountability structures are only meaningful when they carry real consequences and real authority. The individual accountable for an AI system's operational performance must have the authority to modify that system's configuration, escalate concerns about its behavior, and initiate remediation when performance deviates from requirements. Accountability without authority is a nominal assignment that will not produce the behavior governance requires. Similarly, the individual accountable for governance must have the authority to place a system on hold, require additional documentation, or escalate to executive leadership when governance requirements are not met. Accountability structures that cannot say no to the operational teams deploying AI systems are governance in name only.

Diagram 7.2: Three-Level Accountability Structure for AI Security Systems

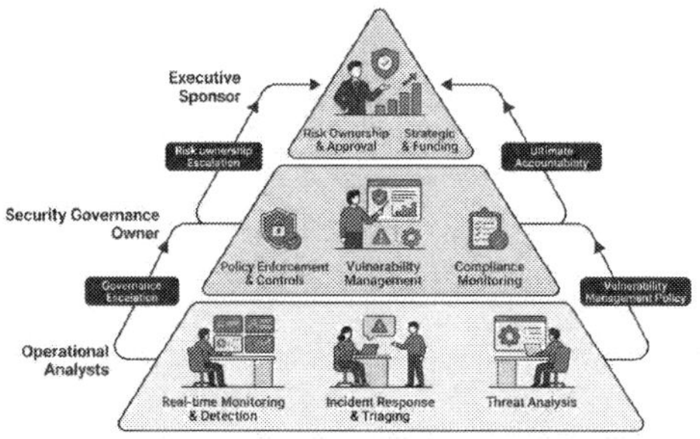

Clearly defined roles ensure proactive management and timely escalation of security risks within AI systems.

7.3 Responsible Use Frameworks for Defensive AI

Defining the Boundary Between Automated and Human-Authorized Action

Every AI security system performs a mix of automated actions and human-assisted actions. The boundary between them, where the machine acts without waiting for a human, and where it pauses and queues for review, is one of the most consequential design decisions an organization makes. That boundary is rarely discussed as a governance decision. It is more often set implicitly, through default vendor configurations, through analyst workload constraints, or through ad hoc adjustments made during incident response. Each of those implicit adjustments expands or contracts the scope of automated authority without organizational review. Over time, the accumulated effect of small, invisible boundary changes can transform a system designed as decision support into one that makes consequential decisions with no human in the loop.

A responsible use framework makes the automation boundary explicit. It specifies, by action type and confidence threshold, which AI outputs trigger automated responses and which trigger human notification. Network isolation of a compromised endpoint may be fully automated when the confidence score exceeds 95%. Account lockout for a suspected credential theft may require analyst confirmation at any confidence level. The framework documents the rationale for each boundary, the evidence that supports the threshold, and the review cadence for revisiting those decisions. When the framework is documented and owned by a named individual, the automation boundary becomes a governance artifact rather than a configuration default, something that can be evaluated, challenged, and deliberately adjusted rather than simply existing.

Data Collection, Retention, and Use Policies for AI Training Environments

AI security systems require large volumes of behavioral and operational data to function, including network flows, authentication logs, endpoint telemetry, user activity records, and communication metadata. The legal and ethical constraints on that data do not disappear when it enters an AI pipeline. Data collected under one legal basis cannot be repurposed for another without review. Data retained beyond its operational necessity creates both privacy exposure and discovery risk. Data fed into a vendor-managed training environment may be used to improve models that benefit the vendor's other customers, a consideration that requires explicit contractual attention. Organizations that have not mapped the data flows of their AI security systems against applicable legal requirements are carrying compliance exposure they may not have quantified.

Organizations should maintain a data inventory that maps each AI system to the data categories it processes, the legal basis for collection, the retention schedule, and the parties with access. That inventory should be reviewed whenever a system's configuration changes, a new data source is connected, or a vendor relationship is modified. In regulated industries, the inventory itself may be a compliance obligation. In all industries, it is the minimum responsible practice for organizations that have agreed, implicitly or explicitly, to protect the privacy of the people whose data their AI systems consume. The data inventory is also the foundation for data subject rights management; organizations cannot respond to access, deletion, or correction requests if they do not know which systems hold relevant data.

Transparency Obligations to Employees, Customers, and Partners

Transparency about AI security monitoring does not require disclosing operational details that would help adversaries evade detection. It requires that people whose behavior is monitored understand, at a general level, that monitoring occurs, what categories of data are collected, and what consequences may follow from automated determinations. In many jurisdictions, this is already a legal requirement. In all organizations, it is a foundation of trust. Employees who discover that AI systems have been monitoring their behavior without disclosure respond with a combination of betrayal and suspicion, which damages the security program's credibility and the broader employment relationship. The appropriate level of transparency is not zero detail, that is, concealment, nor is it a full technical specification, which creates evasion risk. It is an honest general disclosure sufficient to provide the affected parties with meaningful notice.

For customers and partners, transparency obligations extend to the consequences of AI security determinations for their interactions with the organization. If an AI system restricts a customer's account access based on behavioral analysis, the customer deserves to know that the restriction occurred and to have a meaningful path to challenge it. If a partner's systems are subject to AI-powered monitoring as a condition of accessing the organization's infrastructure, that condition should be disclosed in the partnership agreement. These are not merely legal requirements; they are the practices that sustain the trust on which the organization's external relationships depend.

Diagram 7.3: Responsible Use Framework Decision Matrix

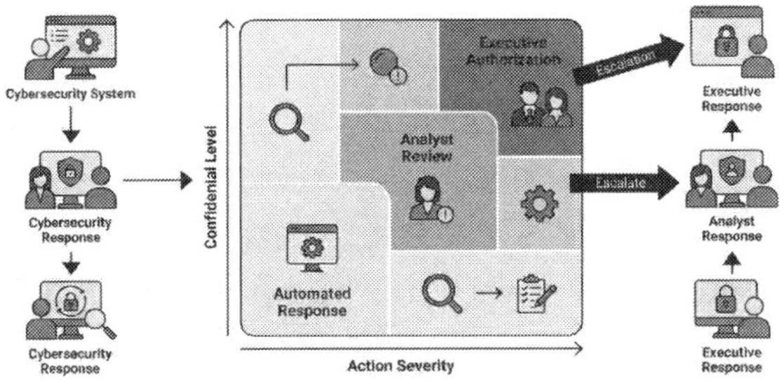

A standardized matrix for determining the appropriate level of response for cybersecurity incidents based on risk and certainty.

7.4 Regulatory Landscape and Compliance Strategy

Current and Emerging AI Security Regulations Across Key Jurisdictions

The regulatory environment for AI in security contexts is fragmenting rapidly, with different jurisdictions moving at different speeds and through different mechanisms. The European Union's AI Act classifies certain security-related AI applications as high-risk, requiring conformity assessments, technical documentation, human oversight provisions, and registration in an EU database before deployment. The act's requirements for transparency, accuracy, and robustness apply directly to AI systems used to protect critical infrastructure. United States federal regulation remains fragmented by sector. Still, the National Institute of Standards and Technology's AI Risk Management Framework has become a de facto standard for organizations seeking to demonstrate responsible AI governance, and OMB guidance increasingly requires federal agencies to apply it to all AI systems that affect individuals' rights or safety.

In the financial sector, the Office of the Comptroller of the Currency, the Federal Reserve, and the FDIC have each issued guidance extending model risk management principles to AI systems. Healthcare organizations face HIPAA scrutiny when AI systems process protected health information, and the FDA has begun regulating AI-enabled clinical decision support systems. Critical infrastructure operators face sector-specific requirements from CISA, NERC CIP, and other regulatory bodies that are actively updating their frameworks to address AI. The practical implication for security managers is that no single compliance framework covers every jurisdiction and sector. A compliance strategy that works requires mapping the organization's specific AI deployments against the specific requirements of every applicable regulatory regime, a mapping exercise that itself requires dedicated resources to perform and to maintain as requirements evolve.

Sector-Specific Requirements: Finance, Healthcare, Critical Infrastructure, Federal

Financial services firms deploying AI for fraud detection, transaction monitoring, or credit-related decisions face model risk management expectations developed over years of regulatory guidance. SR 11-7, originally issued for traditional statistical models, has been explicitly extended by regulators to cover machine learning. Banks must validate AI models before deployment, document their performance over time, and maintain the capability to explain model outputs to examiners. For security-specific AI applications, insider threat detection, and anomaly-based fraud flagging, the same expectations apply. Security teams in financial institutions should expect their AI security tools to be treated as models subject to model risk governance, not as operational software exempt from those requirements.

Healthcare organizations face distinct pressures. AI systems that access or process protected health information are subject to HIPAA's minimum necessary standard, security rule requirements, and business associate agreement obligations. When AI security monitoring tools process clinical data as they often must, given that clinical systems are prime attack targets, the intersection of security necessity and privacy obligation requires careful navigation. Federal agencies face the most prescriptive requirements, with OMB guidance establishing specific requirements for AI use cases that affect rights or safety, including documentation, testing, and ongoing monitoring obligations that security teams must incorporate into their program management practices. Sector-specific requirements will continue to tighten; the organizations that build compliance into operational practice now will not need to scramble later.

Building Audit-Ready Documentation for AI Security Systems

Audit-ready documentation is not produced during the audit. It is maintained continuously as an operational output of the AI security program. The minimum documentation set for each AI security system includes: a system description covering its purpose, architecture, and data inputs; a risk assessment addressing potential harms and mitigations; performance metrics tracked over time; incident records for any failures or unexpected behaviors; configuration change history; vendor assessment records; and approval documentation for deployment and any significant modifications. This documentation should be version-controlled, owned by a named individual, and reviewed at least annually or whenever a material change occurs. Documentation that is not actively maintained degrades quickly. A system description that was accurate at deployment and has not been updated through three major configuration changes is worse than useless because it gives auditors a false picture.

Diagram 7.4: AI Security System Documentation Architecture

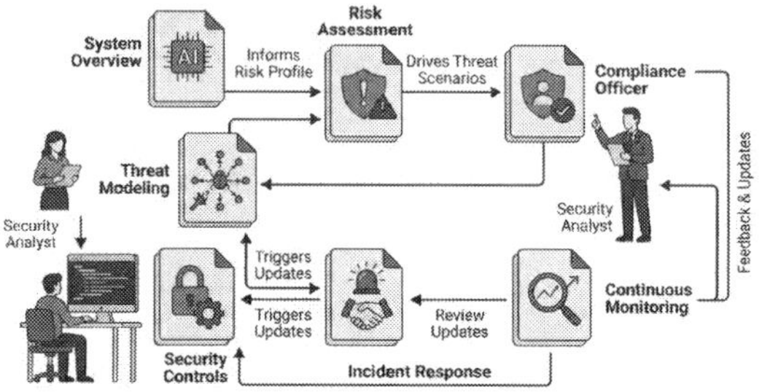

Ensuring comprehensive security documentation for deployed AI systems requires continuous review and iterative updates across core architectural components.

7.5 Governance Architecture for AI Security Programs

Board and Executive Oversight Structures for AI Security Risk

Boards and executive teams cannot manage what they cannot see. AI security risk requires dedicated visibility at the governance level, not because board members will evaluate model architecture or review training datasets, but because AI security risk has strategic implications that require executive judgment: how much automated authority is appropriate, what legal exposure the organization is willing to accept, how AI-related incidents will be disclosed, and how quickly the organization can respond when a core security system produces a damaging outcome. These are not technical questions. They are risk appetite questions that only executive leadership can answer, and they cannot be answered by executives who have not been given the information needed to understand what is at stake.

Effective board-level governance structures designate a specific executive, typically the CISO, CTO, or Chief Risk Officer, as accountable for AI security risk. They include AI-related items in the standing agenda for audit or risk committee meetings, not as standalone briefings but as a regular component of enterprise risk reporting. They require management to report on AI security incidents, near-misses, and significant configuration changes using a consistent framework that allows the board to track trends over time. And they incorporate AI risk into the enterprise risk register, using the same structure as for other material risks: likelihood, potential impact, current mitigations, and residual exposure. Without this structure, AI security risk remains invisible at the level where resource allocation decisions are made.

Policy Frameworks: From Acceptable Use to Incident Classification

An AI security governance program requires a policy architecture that addresses the full lifecycle of AI use in security operations. An acceptable use policy defines which AI applications are authorized in security contexts, under what conditions, and by whom. A data governance policy governs the collection, processing, retention, and sharing of data used in AI systems. An incident classification policy defines what constitutes an AI-related security incident, including incidents caused by AI system failures, not just incidents detected or responded to by AI systems. It establishes the reporting and response obligations set forth below. A vendor governance policy establishes due diligence, contractual, and ongoing oversight requirements for AI tools procured from third parties. Each of these policies must be aligned with the others; gaps or contradictions between them create ambiguity that practitioners will resolve in the least restrictive direction.

These policies must be written with sufficient specificity to be operational rather than merely aspirational. A policy requiring AI systems to be used responsibly has no practical effect. A policy that requires documented human review before any AI-generated alert is elevated to a critical incident, prohibits AI systems from executing network isolation without analyst confirmation, except for a defined set of high-confidence ransomware indicators, and mandates quarterly review of automation thresholds that shape policy behavior and create accountability. The gap between general and specific is where most organizational AI governance frameworks fail. General policies signal intent; specific policies drive behavior.

Third-Party AI Governance: Vendor Assessment and Contractual Controls

The majority of AI security capabilities organizations deploy are purchased, not built. Vendor-managed AI systems present governance challenges that internal systems do not: the organization cannot inspect the model, cannot audit the training data, and cannot access the full range of performance metrics it would generate for an internally developed system. Vendor governance for AI security tools requires a structured assessment that goes beyond standard security questionnaires. It must address the vendor's explainability capabilities, its model update and notification practices, its data handling obligations, its incident reporting requirements, and its approach to testing for bias and adversarial robustness. Vendors who cannot answer these questions with documented evidence rather than sales narratives deserve scrutiny before deployment, not after.

Contractual controls must enforce governance requirements, not merely request them. Contracts should specify performance standards with remediation obligations. They should require advance notification of material model changes, with a review period before changes take effect in the customer's environment. They should prohibit the vendor from using customer data to train models that benefit other customers without explicit consent. They should include audit rights that allow the customer to verify compliance with documentation and testing obligations. These are negotiating points, and vendors vary in their willingness to accept them. Still, an organization that cannot obtain these commitments is accepting a level of governance opacity that may not survive regulatory scrutiny. The negotiation investment is worth making.

Diagram 7.5: AI Vendor Governance Assessment Framework

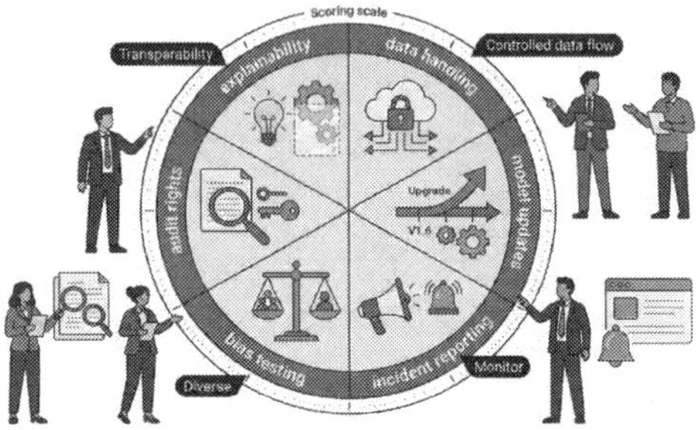

A structured approach to evaluating AI vendor security and compliance.

7.6 Manager's Governance Checklist

Inventory every AI security system in operation, including vendor-managed tools, and document for each: its purpose, the data it processes, the automated actions it is authorized to take, the explainability mechanisms it provides, and the individual accountable for its governance. Where that documentation does not exist, treat the gap as a risk item requiring immediate remediation, not a future project. Documentation gaps discovered during a regulatory examination are significantly more costly to address than documentation gaps discovered through an internal audit.

Define the automation boundary for every AI security system in writing. Specify which outputs trigger automated actions, which require analyst confirmation, and which require executive authorization. Document the rationale for each threshold. Establish a review cadence, at a minimum, annually and following any significant incident involving the system, to revisit those boundaries as model performance and the threat environment evolve. Automation boundaries that are not formally reviewed tend to expand through operational convenience rather than deliberate governance decisions.

Conduct a regulatory mapping exercise to identify all applicable jurisdictional and sectoral frameworks for the organization's AI security deployments. Assign compliance ownership for each requirement. Build the documentation practices that each regime requires into normal operational workflows, so that audit preparation is a continuous activity rather than a reactive scramble. Assign a specific person to track regulatory developments in each applicable jurisdiction and to brief the governance committee on changes that require program response.

Review all vendor contracts for AI security tools against the governance requirements identified in this chapter. Where contracts lack required provisions, notification of model changes, audit rights, and data use restrictions initiates renegotiation. Where renegotiation is not possible, document the residual risk and ensure it appears in the enterprise risk register with appropriate executive visibility. Do not allow contract renewal cycles to pass without addressing governance gaps in vendor agreements.

Establish a cross-functional AI governance committee that includes security, legal, compliance, privacy, and business representation. Assign it a formal mandate, a meeting cadence, and decision rights over AI-related policy questions. Ensure it has a clear escalation path to executive leadership for decisions that exceed its authority. Governance structures that exist only on paper produce governance outcomes that exist only on paper. The committee's effectiveness is determined by the quality of the decisions it makes and acts on, not by the frequency of its meetings.

Diagram 7.6: AI Security Governance Operating Model

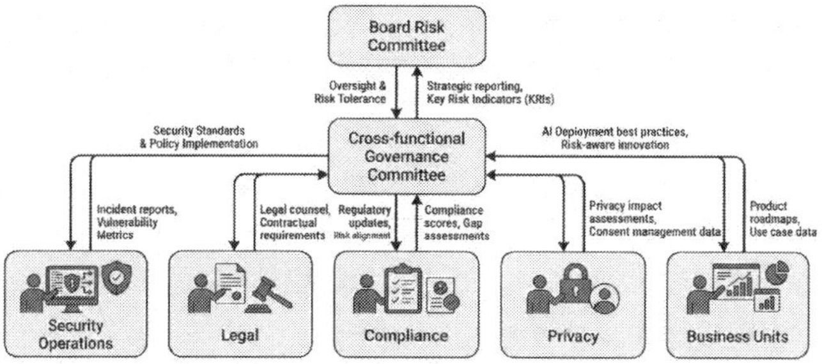

An established AI security governance structure ensures ethical, compliant, and risk-aware deployment of AI technologies across the enterprise.

7.7 The Governance Imperative

The scenario at the start of this chapter, a CISO unable to document how her AI security tools made decisions, will become more common before it becomes less common. The regulatory environment is accelerating, AI capabilities are expanding, and the distance between what organizations have deployed and what they can explain is widening. The organizations that close that distance will not do so by deploying more technology. They will do so by building governance structures that are as rigorous as their technical implementations. Technology governance is not the opposite of technology adoption. It is the precondition for technology adoption that can be sustained.

Governing AI security systems requires the same discipline that effective security programs apply to everything else: clear ownership, documented processes, measured performance, and continuous improvement. It requires extending that discipline to third parties whose tools make consequential decisions inside the organization's environment. And it requires making governance a standing agenda item for leadership, not a project that happens before an audit, but a function that runs every day the AI systems are operating. Organizations that build this discipline will not just satisfy regulators. They will operate AI security programs that they actually understand and can improve.

The AI security program that cannot answer the inspector general's questions is not merely a compliance problem. It is an accountability, a trust, and a risk management problem. Governance is the mechanism that converts accountability from an aspiration into an operational reality. Building it before it is demanded is the only responsible approach.

8 The Team That Security Now Requires

The critical infrastructure operator had spent 18 months deploying an AI-powered threat-detection platform. The technology performed as the vendor demonstrated. The problem was the team. The organization's most experienced security engineers, people who had spent careers building and operating industrial control system defenses, could not interpret the platform's probabilistic risk scores with confidence, could not evaluate whether the model's behavioral baselines were correctly calibrated, and could not write the integration logic needed to connect the AI system to the organization's legacy SCADA monitoring infrastructure. When the vendor's professional services team completed the deployment engagement and left, the internal team was operating a sophisticated AI system with the theoretical understanding of someone who had attended a three-day product training. Incidents were still being caught. But the organization's ability to improve, configure, or critically evaluate the system had not advanced at all. The technology has been acquired. The capability had not.

8.1 Why the Human Dimension Determines AI Security Outcomes

Organizations routinely underinvest in the human dimension of AI security capability. The procurement process centers on technology features rather than on the organizational competencies required to use them effectively. The implementation process focuses on technical deployment rather than on building the understanding needed for the team to adapt the system to changing conditions. The result is a pattern that repeats across industries: sophisticated AI security tools are deployed and then managed in ways that capture only a fraction of their potential value, because the teams operating them have not developed the skills to go beyond the vendor's default configuration. The marginal return on additional technology investment diminishes rapidly when the team cannot fully exploit the technology already deployed.

The stakes of this underinvestment are rising. As AI systems make more consequential decisions, suppressing alerts, isolating endpoints, and restricting access, the quality of human judgment applied to configuring, calibrating, and overriding those systems matters more, not less. The analyst who cannot explain why a behavioral analytics model is producing elevated scores for a particular user group cannot determine whether the scores reflect genuine risk or systematic bias. The security engineer who does not understand how an anomaly detection model was trained cannot assess whether it is appropriate for the organization's specific environment. Capability gaps that were tolerable when AI was an enhancement to human judgment become critical when AI becomes a primary decision-making mechanism with the authority to take actions that are difficult to reverse.

The pace of change compounds the workforce challenge. The skills relevant to AI security today are not the same skills that will be relevant in three years. Machine learning techniques have evolved. New attack categories emerge, requiring new defensive capabilities. AI system architectures change in ways that require updated operational knowledge. Security organizations that adopt a static model of workforce development, train once, deploy indefinitely, will find their teams falling behind technology faster than any single training program can keep pace. The organizations that build continuous learning into their workforce model will maintain the human capability needed to keep pace with a technology environment that does not offer the option of standing still.

Diagram 8.1: The AI Security Capability Gap: Technology Deployment Versus Team Readiness

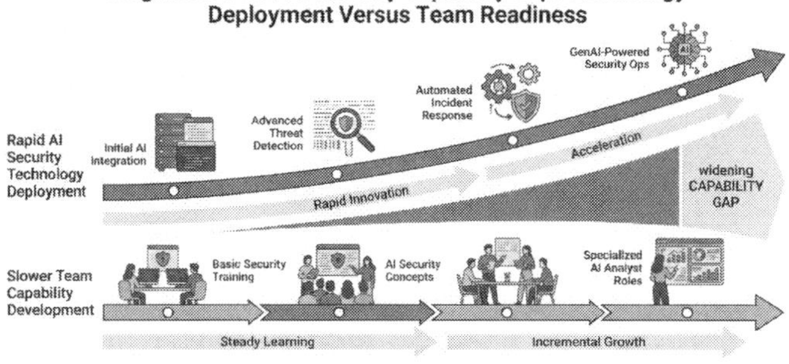

The rapid acceleration of AI security tools outpaces existing workforce skills, creating a critical **capability gap** that requires targeted training and investment.

8.2 Security Team Evolution and Skill Architecture

The Competency Landscape: What AI-Era Security Professionals Need

The competency requirements for security professionals in AI-augmented organizations span a broader range than those demanded of traditional security roles. At the foundation, practitioners still need the domain knowledge that has always defined the field: network security principles, endpoint security, identity and access management, incident response methodology, and threat intelligence interpretation. AI does not replace these; they are the context within which AI outputs must be interpreted. A behavioral analytics alert means nothing to an analyst who does not understand what normal network behavior looks like. An AI-generated attribution of a threat actor is worthless to a responder who cannot evaluate the quality of the underlying evidence.

Above the security foundation, AI-era practitioners need an AI literacy layer to work effectively alongside automated systems. This layer includes conceptual understanding of how common machine learning techniques work, supervised versus unsupervised learning, the role of training data, and the concept of model drift without requiring the mathematical depth of a data scientist. It includes practical skills for interpreting model output, evaluating confidence indicators, and recognizing failure modes that lead to incorrect results. And it includes the judgment to know when to follow an AI recommendation, when to seek additional information before acting, and when to override the system entirely. This judgment cannot be codified into a rule; it requires both technical understanding and operational experience to develop.

A third competency layer, required for more senior roles, is the ability to evaluate AI systems themselves: to assess whether a vendor's model's claims are credible, to design meaningful tests of system performance, to identify data quality problems that may degrade model outputs, and to contribute to governance discussions about automation boundaries and accountability structures. This layer requires deeper AI knowledge but does not require the ability to build or train models from scratch. Organizations that conflate AI practitioners with AI security practitioners, assuming that hiring data scientists will solve their operational security capability gap, will discover this conflation through operational failures. The gap is between AI knowledge in the abstract and AI knowledge applied to security-specific operational problems.

Data Science and Machine Learning Literacy for Security Practitioners

Machine learning literacy for security practitioners is a well-defined, teachable body of knowledge distinct from data science as a profession. The core curriculum includes: how classification models learn to distinguish between normal and anomalous behavior; how training data quality affects model performance; what false positive and false negative rates mean in operational terms and how they trade off against each other; how feature engineering translates raw security telemetry into inputs a model can process; and how model drift occurs as the environment the model was trained on diverges from the environment it is operating in. Each of these concepts has direct operational implications for security practitioners to apply in their daily work.

These concepts are learnable by experienced security practitioners in weeks, not years. The challenge is not the intellectual difficulty; it is the absence of structured learning pathways that present these concepts in a security context. Generic data science training teaches these ideas using business analytics examples that do not transfer well to the mental models security practitioners bring to their work. Security-specific AI literacy programs that use network anomaly detection, insider threat modeling, and malware classification as their teaching examples produce significantly faster and more durable learning. The curriculum does not need to be long. A focused twelve-week program covering the concepts most relevant to the tools the team operates can produce meaningful capability improvement across an entire analyst cohort.

Role Design: New Titles, Responsibilities, and Career Pathways

The emergence of AI in security operations has created genuine new roles that existing job families do not cleanly accommodate. AI Security Engineer roles sit at the intersection of security engineering and machine learning engineering, responsible for the technical implementation and maintenance of AI security systems, including model integration, data pipeline management, and performance monitoring. Threat Intelligence Analysts with AI specialization translate raw intelligence into inputs that AI systems can act upon and interpret AI-generated intelligence outputs in the context of a broader threat picture. AI Security Governance Specialists own the documentation, compliance, and oversight structures that keep AI deployments within authorized boundaries. Each of these roles requires a combination of competencies that traditional job descriptions were not designed to capture.

Career pathways for these roles require intentional design. An experienced analyst who wants to develop an AI security specialization needs a progression that builds on existing domain expertise rather than requiring a lateral career move into data science. A security engineer who wants to move into AI governance needs access to legal and compliance knowledge that traditional engineering paths do not include. Organizations that design these pathways explicitly with defined competency milestones, learning resources, and progression criteria retain the experienced practitioners who might otherwise leave for organizations that offer clearer development opportunities. The organizations most likely to lose experienced practitioners to the talent market are those that introduce AI without explaining how it changes career trajectories.

Diagram 8.2: AI Security Role Architecture and Career Pathways

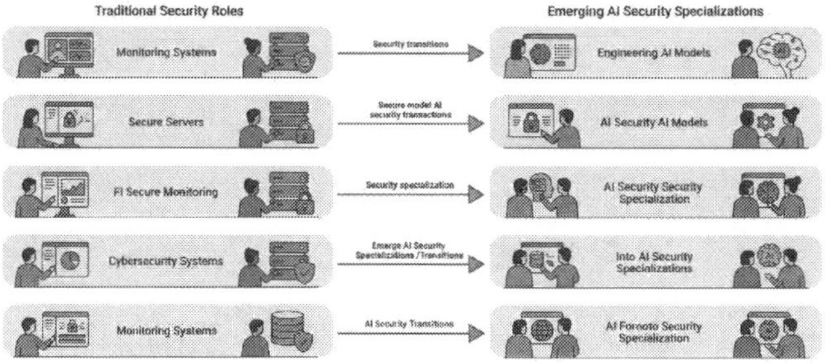

The growing integration of Artificial Intelligence creates new pathways from traditional cybersecurity roles into specialized AI security disciplines.

8.3 AI Skills Development and Talent Strategy

Building Internal Capability Versus Strategic Partnership and Acquisition

The build-versus-buy decision for AI security capabilities takes a different form than the same question for AI security technology. Technology can be turned on quickly and turned off if it underperforms. The organizational knowledge needed to use technology effectively takes time to develop and is quickly lost when experienced practitioners leave. Organizations that rely entirely on vendor capability to operate their AI security tools have not built a security program. They have outsourced one. Dependency is expensive, difficult to manage, and creates single points of failure at critical moments. The question is not whether to build internal capability, but which capabilities to build and which to supplement through strategic partnership.

Internal capability investment should focus on the skills that provide the most strategic leverage: critically evaluating AI security tools, configuring and calibrating them for the organization's specific environment, interpreting their outputs in an operational context, and governing their use within appropriate boundaries. These capabilities enable an organization to derive differentiated value from commodity technology. Deep model development, the ability to build and train custom AI security models from scratch, is rarely a core competency that security organizations need to develop internally. It can be more efficiently accessed through academic partnerships, specialized vendors, or government research programs. The organization's energy is better spent on mastering the available tools than on building tools that the market already provides.

Learning Programs That Keep Pace with AI Capability Development

The half-life of AI security knowledge is short. A learning program designed around today's AI security capabilities will be partially obsolete within eighteen months. Organizations that invest heavily in one-time training programs and then declare the capability gap closed will find themselves back at the same gap in two years. Effective learning programs for AI security competency are designed as ongoing systems rather than one-time events. They include structured technical training updated at least annually, access to practitioner communities where emerging techniques and tools are discussed, internal knowledge-sharing mechanisms that distribute learning from individual practitioners to the broader team, and protected time for practitioners to stay current with developments in the field. Each of these components is necessary; a program that includes structured training but provides no protected time for self-directed learning will see practitioners apply training in the first month and then fall behind as their operational schedule crowds out further development.

Effective programs also connect learning to operational practice. Training that teaches AI security concepts in a classroom setting without connecting those concepts to the specific tools and workflows the practitioner encounters every day produces knowledge that does not transfer to job performance. The most effective learning programs embed practice within the operational environment: table-top exercises that require analysts to evaluate AI-generated scenarios, red team activities that probe AI system failure modes, and structured after-action reviews that use incident data to surface what the AI got right, what it missed, and why. When learning and operational practice are integrated, practitioners develop the contextual judgment that classroom learning alone cannot produce.

Recruiting for AI-Security Hybrid Roles in a Competitive Market

The market for professionals who combine genuine security domain expertise with meaningful AI literacy is genuinely competitive. These individuals are sought simultaneously by technology vendors, financial institutions, consulting firms, and government agencies. Organizations that cannot offer competitive compensation, meaningful technical challenges, clear development opportunities, and reasonable working conditions will lose recruiting competitions to those that can. Mission-driven organizations, government agencies, critical infrastructure operators, and healthcare systems have non-compensation advantages they frequently fail to use effectively: the significance of the work, the breadth of technical exposure, and the stability of employment. Communicating these advantages clearly and credibly in recruiting processes is a differentiating capability that most mission-driven organizations underuse.

An alternative recruiting strategy is to hire for one of the two required competencies and develop the other. Security practitioners who understand the domain can be developed into AI-literate practitioners through structured learning programs. Data scientists who understand machine learning can be developed into security-aware practitioners through mentorship and operational exposure. Both development paths take 12 to 24 months to achieve full effectiveness. Organizations with patience and a commitment to developing investment will build a more loyal and more deeply capable team than those who chase fully formed hybrids in a market that does not produce them in sufficient numbers. The development investment also signals to the team that the organization views practitioners as worth investing in, thereby improving retention.

Diagram 8.3: AI Security Talent Strategy Matrix

A strategic framework for optimizing organizational cybersecurity capabilities through tailored AI talent acquisition and development.

8.4 Human-AI Teaming Models

Workflow Design for Effective Human-Machine Collaboration

Human-AI teaming in security operations does not emerge spontaneously from the deployment of AI tools alongside human analysts. It requires deliberate workflow design that assigns tasks to humans and machines based on their respective strengths, structures handoff points between automated and human actions, and creates feedback mechanisms that enable both sides of the collaboration to improve over time. The default in most organizations is that AI generates outputs, and humans react to them, which is not human-AI teaming. It is AI-driven operations with a human safety net. This structure tends to overwhelm human judgment with volume, underuse human expertise in pattern recognition and contextual interpretation, and provide no systematic mechanism for human insight to improve AI performance over time.

Effective workflow design starts with the question: what can humans do that AI cannot, and vice versa? AI systems excel at rapidly processing large volumes of telemetry, maintaining consistent attention over extended time periods, and applying learned pattern recognition without the variability that fatigue introduces into human attention. Human analysts excel at contextual reasoning, creative hypothesis generation, interpretation of ambiguous or novel situations, and the kind of institutional knowledge that is not available in training data. Workflows that route volume-intensive, pattern-matching tasks to AI and reserve contextual judgment, novel situation assessment, and high-stakes authorization decisions for humans are using both capabilities where they are strongest. Workflows that require humans to review every AI output at high volume are using neither capability well.

Cognitive Load Management When Working Alongside AI Systems

Working alongside AI systems creates cognitive load patterns that differ from traditional security operations. Analysts must evaluate AI outputs rather than generating assessments from raw data, a shift that sounds like it reduces cognitive effort but in practice introduces new demands: evaluating whether the AI's reasoning is sound, maintaining awareness of the AI system's known failure modes, and sustaining the domain expertise needed to catch errors that the AI makes with high confidence. The risk of automation bias, the tendency to accept AI outputs without sufficient critical evaluation, is real and documented across multiple high-stakes domains, including aviation, medicine, and intelligence analysis. Organizations cannot assume that experienced practitioners are immune to automation bias; research consistently shows that expertise reduces but does not eliminate the tendency.

Organizations should design analyst workflows to counteract automation bias structurally, not rely on individual vigilance to overcome it. This means requiring analysts to record their own hypothesis about an alert before viewing the AI's assessment, not after. It means rotating analysts between AI-assisted and traditional analysis tasks to preserve manual skills that atrophy with disuse. It means creating a culture where disagreement with AI outputs is treated as a valuable signal, not as a failure of the analyst to trust the technology. The organizations that manage cognitive load well will sustain the quality of human judgment that gives the human-AI collaboration its resilience. Those who do not will gradually accumulate a hidden competency deficit as manual skills decay, assuming the AI handles those assessments now.

Trust Calibration: Knowing When to Follow and When to Override AI

Trust calibration, the degree to which an analyst's confidence in AI outputs matches those outputs' actual reliability, is a measurable organizational capability. Under-calibrated analysts ignore AI recommendations even when they are correct, losing the efficiency benefits of AI augmentation. Over-calibrated analysts follow AI recommendations even when they are wrong, losing the safety benefits of human oversight. Neither failure mode is obvious in real-time; both are detectable through outcome tracking over time. Organizations should measure calibration systematically: when analysts override AI recommendations, what fraction of those overrides are correct? When analysts follow AI recommendations without modification, what fraction of those turn out to be wrong? These metrics are available in any organization that instruments its incident management process with sufficient fidelity to track the outcomes of individual analyst decisions.

Calibration training is distinct from AI literacy training. It does not focus on understanding how AI systems work; it focuses on developing accurate intuitions about when specific AI systems are likely to be right or wrong. This requires exposure to the AI system's actual performance history, not just aggregate statistics, but specific cases where the system failed and specific patterns that predict failure. Analysts who have seen cases in which the system confidently produced an incorrect output develop more calibrated skepticism than analysts who have only been told that false positives occur at a given rate. Case libraries of AI system failures are among the most valuable training resources an organization can build, and they are entirely within the organization's ability to construct from its own operational data.

Diagram 8.4: Human-AI Trust Calibration Spectrum

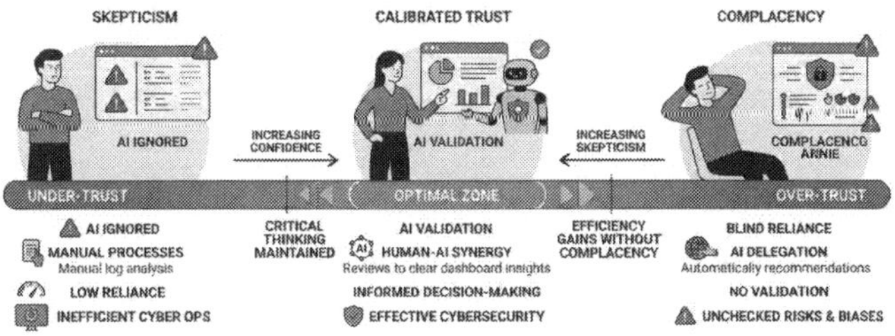

Calibrating human trust in AI systems ensures balanced collaboration, leveraging automated insights while maintaining **critical human judgment** in complex cybersecurity operations.

8.5 Organizational Culture and Change Management

Leading the Transition: Addressing Anxiety About AI-Driven Role Change

The introduction of AI into security operations generates genuine anxiety among experienced practitioners. Some of that anxiety is about job security, the fear that AI will eliminate positions and reduce the value of years of accumulated expertise. Some is about competence, the fear of not being able to operate effectively in an AI-augmented environment and being judged inadequate by standards that changed without warning. Some is about loss of control, the discomfort of working in an environment where systems make consequential decisions that the practitioner cannot fully inspect or understand. Leaders who dismiss these anxieties as irrational will fail to address them, and unaddressed anxiety produces resistance that undermines AI adoption in ways that are difficult to diagnose and expensive to reverse.

Addressing anxiety requires honesty about what AI is and is not changing. Job elimination from AI in security operations is less common than job transformation. The analyst who spent most of her shift triaging low-confidence alerts will spend more of her shift on investigation and response work that requires genuine expertise if the organization is intentional about redesigning the role alongside the technology. The engineer who built detection rules manually will shift toward evaluating and improving AI-generated detection logic if the organization invests in the transition. Leaders who can articulate this transformation concretely, with specific examples relevant to their team's work, will find that anxiety about the unknown gives way to constructive engagement with manageable change. Leaders who offer only vague reassurances will find that anxiety persists and hardens into resistance.

Building a Culture of Continuous Learning in Security Organizations

Security organizations have traditionally valued deep expertise in established domains. The same expertise that makes an experienced practitioner effective can make continuous learning feel like a threat, an implicit message that what the practitioner knows is no longer enough. Building a culture of continuous learning requires reframing: not an acknowledgment that existing knowledge is obsolete, but an expectation that adding new knowledge to an existing foundation is a normal part of professional practice, as it is in medicine, law, and engineering. Leaders model this expectation by learning visibly, by sharing what they are reading and thinking about, and by treating the admission of uncertainty as a sign of intellectual seriousness rather than professional vulnerability.

Structural mechanisms reinforce the culture. Protected time for professional development, not aspirationally allocated but actually defended against operational pressure, sends the message that learning is a legitimate use of working hours. Internal knowledge-sharing forums where practitioners present what they have learned from recent incidents, research, or training convert individual learning into organizational capability. Recognition systems that reward learning activity, not just operational performance, signal that development is valued alongside execution. Organizations that build these structures and protect them consistently will find that the culture of continuous learning they are trying to create gradually becomes self-sustaining as practitioners experience its benefits firsthand.

Measuring Organizational Readiness for AI-Augmented Security Operations

Readiness for AI-augmented security operations is a measurable state, not a judgment call. Organizations should assess readiness across four dimensions: technical readiness, covering the data infrastructure, integration capabilities, and system maturity needed to support AI deployment; skill readiness, covering the distribution of AI literacy across security roles; process readiness, covering the documented workflows, governance structures, and oversight mechanisms needed to manage AI systems appropriately; and cultural readiness, covering the team's orientation toward continuous learning and their psychological safety in an environment where roles are changing. Each dimension should be assessed with specific, observable indicators, not general impressions. The assessment output should identify specific gaps and owners, not aggregate scores that obscure the detail needed to drive action.

8.6 Manager's Team Development Checklist

Conduct a competency assessment across the security team that maps current AI literacy levels against the requirements for each role. Do not assume that the deployment of AI tools has created AI competency. Use specific, observable criteria: can the analyst explain the difference between a false positive and a false negative in the context of a specific tool?

Can the engineer identify the conditions under which the behavioral analytics model is likely to produce elevated scores for users who are not actually at risk? Document the gaps and assign development plans with timelines.

Design or commission AI literacy training that is security-specific in its examples and context. Evaluate vendor training programs critically: do they teach how to use the specific tool, or do they build a transferable understanding of AI security principles? The former produces tool operators. The latter produces practitioners who can adapt as tools change. Build protected time for training completion into operational schedules rather than leaving it to individual initiative outside working hours.

Review job descriptions and career pathways for AI-era security roles. Identify where AI literacy requirements are absent from job descriptions for roles that require it. Update role profiles, hiring criteria, and performance expectations to reflect the competency requirements of an AI-augmented environment. Communicate the updated expectations to current practitioners with a clear development pathway rather than a sudden change in performance standards.

Design analyst workflows explicitly for human-AI collaboration. Map the handoff points between automated and human action. Build in mechanisms for analyst feedback to reach the teams responsible for AI system configuration and performance. Measure automation bias quarterly by tracking analyst override decisions.

Address workforce anxiety directly in team communications. Specifically, what roles will look like after AI augmentation, what the development pathway is for practitioners whose current work is most affected, and what the organization's commitment to the team will be during the transition. Vague reassurances increase anxiety. Specific plans reduce it. The leaders most effective at managing AI-related workforce anxiety are those who can answer the specific questions their teams are actually asking, not the questions leaders find easier to address.

Diagram 8.5: AI Security Team Readiness Assessment Framework

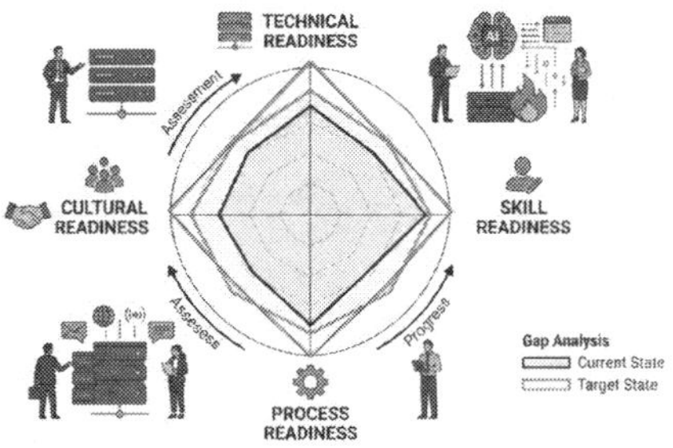

This framework evaluates and identifies critical capability gaps to enhance an organization's overall AI security posture.

8.7 The Human Foundation of AI Security

The critical infrastructure operator in this chapter's opening scenario had the technology. What the organization lacked was the human capability to make the technology perform. That gap is not unusual; it is the default outcome when organizations approach AI security as a technology procurement problem rather than an organizational transformation challenge. Closing it requires investment in learning, role design, workflow architecture, and culture that is as deliberate and sustained as the investment in technology. Organizations that allocate the same rigor to human capability development that they apply to technology selection will extract disproportionate value from identical tool sets.

The specific capabilities the AI era demands of security professionals are learnable. The workflow structures that allow humans and AI systems to complement each other effectively are designable. The cultures that sustain continuous learning are buildable. None of these outcomes happens automatically as a consequence of deploying AI tools. Each requires leadership attention, resource commitment, and organizational patience that spans multiple years.

AI security systems are only as capable as the teams operating them. The teams that develop genuine AI literacy, calibrated trust, and continuous learning habits will extract disproportionate value from the same technology. In contrast, organizations that treat the human dimension as secondary will realize only marginal improvements. The technology advantage is smaller than vendors suggest. The human advantage is larger than most organizations realize.

Diagram 8.6: The AI Security Capability Flywheel

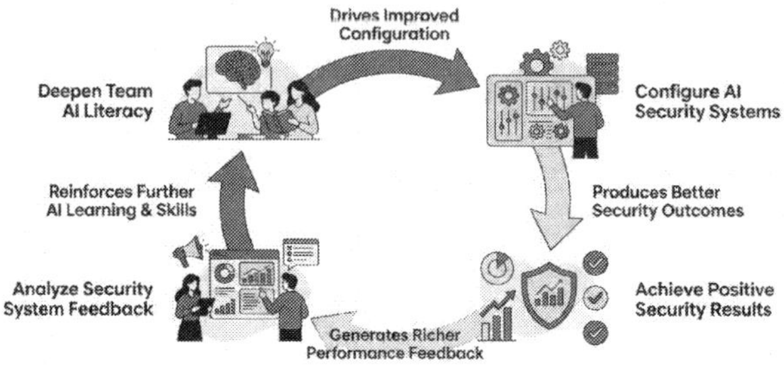

This circular flow demonstrates how improving team understanding of AI security directly enhances system configuration, creating a powerful feedback loop for continuous improvement.

9 Architecture for an Uncertain Future

The multinational corporation's new CISO inherited a five-year security roadmap on her first day. By her third month, she had identified seven major assumptions embedded in that roadmap that developments in AI had already invalidated. The threat landscape the roadmap was designed to address had changed. The defensive tools the roadmap assumed would be deployed were being superseded by capabilities that did not exist when the plan was written. The organizational capabilities the roadmap expected to have built were still aspirational because the skills environment had shifted faster than the training programs could keep pace. She did not revise the roadmap. She replaced it with a different planning architecture: a set of strategic objectives stable enough to guide investment and flexible enough to accommodate change, supported by a continuous assessment process that replaced the annual planning cycle with a quarterly posture review. The plan was not a document. It was practice.

9.1 Why Static Plans Fail in Dynamic Environments

The security planning problem has always included uncertainty. What has changed is the pace and magnitude of the changes that invalidate plans. In the pre-AI era, a five-year security roadmap built on reasonable assumptions about threat evolution and technology development could guide investment decisions with tolerable error. The fundamental nature of threats, the architecture of defensive technology, and the skill requirements for security operations changed slowly enough that a plan written in year one remained mostly relevant in year three. That is no longer true. AI capabilities are advancing at a pace that makes eighteen-month planning cycles feel like long-range strategy. A roadmap written with careful attention to today's AI-augmented threat environment may describe a threat landscape that has materially changed before the first phase of the roadmap is complete.

The failure mode of static planning in dynamic environments is not dramatic. Plans do not become visibly wrong overnight. They become gradually irrelevant: the threat the plan was designed to address is no longer the most significant, but the program continues to invest in that defense because that is what the plan says. The technology the plan assumed would be the organization's primary defensive platform has been superseded, but procurement cycles keep the organization on the old path. The skills the plan assumed would be developed are no longer the right ones, yet training programs continue to deliver the original curriculum. Each of these misalignments is individually tolerable. Together, they produce a security program that works very hard but falls steadily further behind.

The solution is not to plan better. It is to plan differently. Organizations that build planning architectures designed for continuous adaptation with mechanisms to update assumptions, revise investments, and redirect capability development in response to environmental changes maintain strategic coherence without static rigidity. The CISO in this chapter's scenario did not abandon strategy. She replaced a document with practice, trading the false certainty of a five-year plan for the genuine capability to adapt as circumstances change. The organizations that make this transition will be better positioned not just for the current AI-augmented threat environment, but for whatever comes after it.

Diagram 9.1: The Adaptive Planning Cycle Versus the Traditional Annual Roadmap

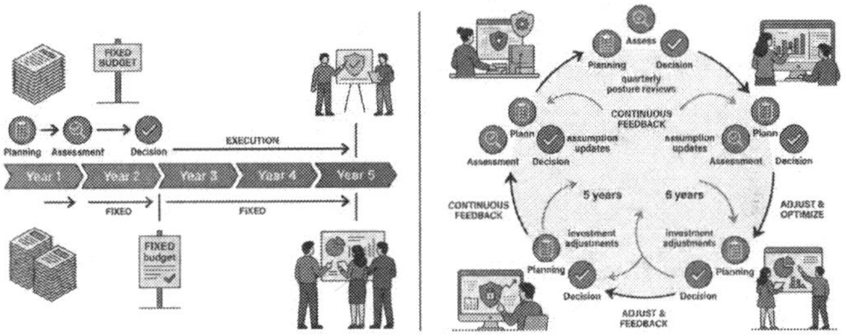

Adaptive planning continuously revises cybersecurity strategies based on real-time data and market shifts, unlike static annual roadmaps.

9.2 Continuous Adaptation as Strategic Discipline

Building Security Programs Designed for Continuous Change

A security program designed for continuous change shares structural characteristics that distinguish it from programs built around stable, long-range plans. Its objectives are expressed in outcomes, not activities: reduce the time from initial compromise to detection, not deploy product X by quarter three. Its investment decisions are made in shorter cycles, with explicit review checkpoints that allow reallocation as priorities shift. Its governance structures include formal mechanisms for surfacing and acting on new information: a quarterly threat environment review, a semi-annual technology assessment, and an annual organizational readiness audit, rather than relying on informal awareness of change to drive strategic adjustment. These mechanisms must be institutionalized with meeting cadences, ownership, and decision rights, not left as aspirational practices.

Modular architecture is a prerequisite for adaptive programs. Security organizations whose tools, processes, and skills are tightly coupled, where changing one requires changing everything, cannot adapt quickly because the cost of change is too high. Programs designed with modularity at their core platform components that can be upgraded or replaced independently, skills that transfer across changing tool sets, processes that are designed around capabilities rather than specific products, can absorb change without the organizational disruption that transformation produces in tightly coupled programs. This is not merely a technology architecture consideration. It applies equally to the security function's organizational design, training program design, and governance framework design. An organization that cannot change its detection engineering workflow without a six-month retraining program cannot respond to a new class of AI-augmented attacks within the required timeframe.

Feedback Loops: From Incident Data to Strategic Investment Decisions

The richest source of information about whether a security program's investments are producing the right outcomes is its incident data. Every security incident contains information about the attack techniques that succeeded, the defensive controls that failed or were absent, the detection mechanisms that identified the incident and those that missed it, and the response capabilities that contained or failed to contain the damage. Organizations that conduct systematic after-action reviews and feed their findings into a structured analysis process can extract strategic investment signals from operational data that most programs treat as closed cases once the immediate remediation is complete.

The analytical question is not what happened in a specific incident, but what the pattern of incidents over time reveals about the alignment between the security program's investments and the actual attack surface. If incidents consistently exploit a category of control the program has underinvested in, that is an investment signal. If detection consistently fails for a category of technique that the program's tools are not calibrated to catch, that is a capability signal. If the response consistently struggles with a class of attack that requires skills the team has not developed, that is a workforce signal. Organizations that institutionalize this analysis through regular reviews of aggregated incident data presented to the appropriate decision-making forum with explicit investment recommendations create a feedback loop that keeps strategic investment aligned with operational reality rather than with the assumptions embedded in last year's plan.

Red Team and Wargaming Programs for Ongoing Posture Assessment

Red team exercises and wargaming programs provide information about security posture that operational metrics cannot. Metrics derived from normal operations tell you how the program performs against the attacks it has been designed to handle. Red team exercises test performance against attacks the program was not specifically designed for, using the full range of techniques a sophisticated adversary might employ, including AI-augmented techniques that the organization may not yet have experienced in production. The gap between operational metrics and red team findings is one of the most useful measures of strategic vulnerability available to security leadership. Organizations that have never conducted a red team exercise do not know the size of that gap.

AI-augmented red team programs produce higher-fidelity assessments because they can operate at a pace and scale that more accurately reflect the adversary environment than manual red teams, constrained by the speed and scope of human action. Organizations should consider building an AI-augmented red team capability, either internally or through specialized vendors, as part of their ongoing posture assessment infrastructure. The objective is not to run a red team exercise once a year and document the findings. It is to maintain a continuous understanding of where the organization's defenses are strongest and most vulnerable, so that strategic investment decisions are informed by current posture data rather than by the posture at the time of the last annual assessment.

Diagram 9.2: Strategic Feedback Loop: Incident Data to Investment Decisions

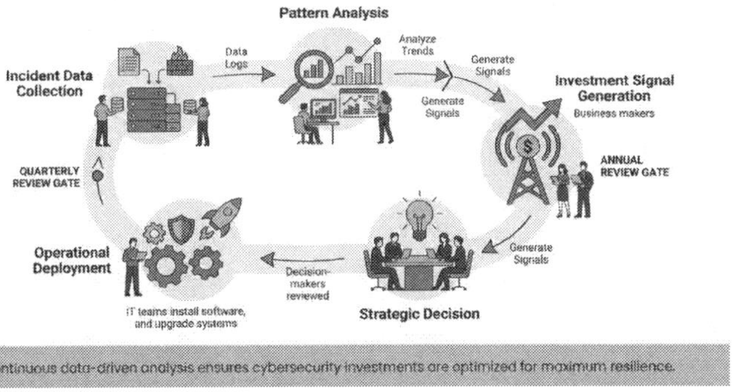

9.3 Threat Intelligence Integration at Enterprise Scale

Operationalizing Intelligence: From Indicators to Strategic Insight

Threat intelligence spans a spectrum from tactical indicators, such as specific IP addresses, file hashes, and domain names associated with known malicious activity, to strategic insight into the goals, capabilities, and targeting patterns of adversary groups. Most organizations operationalize only the tactical end of the spectrum, feeding indicators into detection systems and measuring success by the number of indicators processed. This is a narrow use of intelligence that captures only the lowest-value component of what intelligence programs can provide. Tactical indicators age within days or hours. Strategic insight about adversary intent and capability remains actionable for months or years and informs investment decisions that no amount of tactical indicator processing can address.

Operationalizing strategic intelligence requires connecting the intelligence function to the strategic planning process, not just to the detection engineering team. When intelligence indicates that a threat actor group with capabilities relevant to the organization is actively targeting the sector, that assessment should inform investment priorities, not just generate a new set of detection signatures. When intelligence identifies a new technique that AI-augmented adversaries are using to evade behavioral analytics, that should trigger a review of whether the organization's behavioral analytics platform is configured to detect it, and if not, what the remediation pathway is. The intelligence-to-strategy connection requires both analytical capability and organizational structure: a process for translating intelligence assessments into actionable investment and configuration recommendations that reaches the decision-makers who can act on them.

AI-Powered Intelligence Analysis and Predictive Threat Modeling

The volume of threat intelligence available to enterprise security programs has grown beyond the human capacity to analyze it systematically. OSINT sources, commercial intelligence feeds, government sharing programs, and internal incident data collectively produce a volume of raw intelligence that is impossible to process manually at the speed and consistency required for operational relevance. AI-powered intelligence analysis applies natural language processing, pattern recognition, and predictive modeling to this volume of data, surfacing the most operationally relevant items, identifying patterns across sources that manual analysis would miss, and generating predictive assessments of likely adversary behavior based on observed trends. The analyst's role shifts from data processing to insight validation and contextual interpretation.

Predictive threat modeling using AI to generate probabilistic assessments of where attacks are likely to occur, which techniques adversaries are likely to use, and which organizational assets are most likely to be targeted is an emerging capability that the most sophisticated security programs are beginning to operationalize. It does not replace analyst judgment; it focuses analyst attention on the highest-probability risk scenarios, allowing limited analytical resources to be deployed where they are most likely to matter. Organizations evaluating AI-powered intelligence platforms should look specifically for evidence that predictive capabilities have been validated against historical outcomes, not merely that they exist as a product feature. Predictive models that have never been evaluated against actual events can give a false sense of analytical capability, which can be more dangerous than operating without them.

Information Sharing Ecosystems and Collective Defense Frameworks

No single organization has enough visibility into the threat landscape to maintain an adequate intelligence picture independently. The adversaries that AI has empowered operate simultaneously across multiple organizations, sectors, and geographies. Effective defense requires intelligence sharing at a scale and speed that only collective frameworks can achieve. Information Sharing and Analysis Centers provide sector-specific sharing infrastructure for financial services, healthcare, energy, and other critical sectors. The Cybersecurity and Infrastructure Security Agency operates sharing programs that connect federal government intelligence with private-sector participants. International partnerships extend the collective picture across borders for multinational organizations. Each of these frameworks has different participation requirements, different data standards, and different governance structures that organizations must understand before joining.

The practical barrier to participation is not a lack of willingness but a lack of capacity. Sharing programs require organizations to have the analytical infrastructure to consume intelligence in machine-readable formats, the legal frameworks to share their own incident data without creating liability, and the operational processes to translate shared intelligence into defensive action within operationally relevant timeframes. Organizations that lack any of these elements will participate nominally but not effectively. Building genuine participation in collective defense frameworks requires explicitly addressing each barrier: integrating sharing platform feeds into the SIEM and SOAR environment, developing privacy-preserving anonymization procedures for shared incident data, and assigning analysts with dedicated responsibility for intelligence consumption and contribution. Nominal participation that does not generate or consume actionable intelligence provides no defensive value while consuming the administrative overhead of formal membership.

Diagram 9.3: Threat Intelligence Value Chain: From Raw Data to Strategic Decision

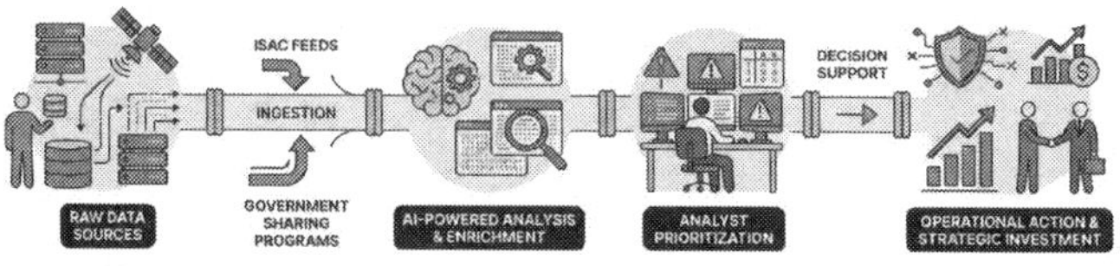

Transforming cybersecurity data streams into prioritized insights for effective operational defense and long-term strategic planning.

9.4 Technology Architecture for Enduring Resilience

Zero Trust Architecture as the Foundation for AI-Era Defense

Zero trust architecture, the design principle that no user, device, or application should be trusted by default, regardless of network location, and that all access must be continuously authenticated and authorized, is not a new idea. It has been articulated as a security design principle for more than fifteen years. What the AI era has changed is the urgency of its implementation. AI-augmented attackers can move through traditionally trusted network segments with unprecedented speed, exploiting the implicit trust that flat network architectures extend to any entity that has cleared the perimeter. The organizations that have implemented zero-trust principles are structurally more resilient to AI-augmented lateral movement than those that have not, because there is no trusted path for a compromised identity to traverse without triggering continuous verification requirements.

Zero-trust implementation is not a technology purchase. It is an architectural journey that requires simultaneous changes to network design, identity and access management, endpoint management, application security, and data governance. Most organizations are at different stages of this journey across domains, with well-advanced identity controls but immature data access governance, or with strong endpoint controls but a flat internal network architecture. The practical approach is to assess zero-trust maturity across each domain, identify the highest-risk gaps, and prioritize remediation where AI-augmented attacks are most likely to exploit implicit trust. CISA's Zero Trust Maturity Model provides a useful framework for structured assessment and prioritization. The organizations that approach zero trust as a destination rather than a journey will repeatedly discover that the work is more extensive than a product evaluation suggested.

Platform Consolidation Versus Best-of-Breed: A Strategic Framework

The platform consolidation-versus-best-of-breed debate takes on a different character in the AI era than in the preceding decade. The argument for consolidation has strengthened: AI security capabilities that operate across an integrated platform, where behavioral analytics, threat intelligence, endpoint detection, and network monitoring share a common data layer, can generate insights that siloed, best-of-breed tools cannot produce. The correlation across data sources that enables AI to detect sophisticated multi-stage attacks requires that those data sources be accessible to the AI system in a consistent, timely format. Siloed data environments constrain the AI's analytical scope to the portion of the attack it can see. AI that can only see network traffic cannot correlate a suspicious login with an unusual file access pattern that occurred thirty minutes later on a different system.

The argument for best-of-breed has not disappeared. Platform vendors whose AI capabilities lag in specific domains create gaps that cannot be addressed without supplementing the platform with specialized tools. The integration overhead of best-of-breed environments has declined as API standards have improved, reducing one of the traditional arguments for consolidation. Organizational dependence on a single platform vendor creates concentration risk, both the operational risk of a single point of failure and the strategic risk of a vendor relationship that is difficult to exit if performance deteriorates or pricing increases beyond the acceptable range. A strategic framework for this decision evaluates the maturity of the platform vendor's AI capabilities in each required domain, the data integration requirements of the AI use cases the organization needs to support, and the acceptable level of vendor concentration risk, with explicit review as vendor capabilities and market alternatives evolve.

Cloud, Edge, and Hybrid Environment Security in an AI-Defined Landscape

The environments that enterprise security programs must protect have expanded beyond the traditional data center perimeter to include public cloud infrastructure, SaaS applications, edge computing environments, and a distributed workforce that accesses systems from an enormous variety of locations and devices. AI security capabilities must extend across this entire footprint, which requires both architectural coherence and consistent visibility and control across environments, and the acceptance that different environments require different security approaches applied consistently toward the same objectives.

Cloud environments present distinctive AI security challenges. The shared responsibility model requires security teams to protect workloads and data in environments where the cloud provider manages the underlying infrastructure. AI security tools must operate within the access and telemetry constraints of each cloud platform's native security services. The dynamic, ephemeral nature of cloud infrastructure workloads that spin up and down, containers that exist for minutes, serverless functions that execute for milliseconds, creates behavioral baselines that are more complex to establish and maintain than those for stable on-premises infrastructure. Organizations should evaluate AI security tools specifically for their native cloud integration capabilities, including support for cloud-native identity and access management services, telemetry collection from cloud platform APIs, and handling of the dynamic infrastructure patterns that cloud environments produce at scale.

Diagram 9.4: Zero Trust Architecture Across Hybrid Environments

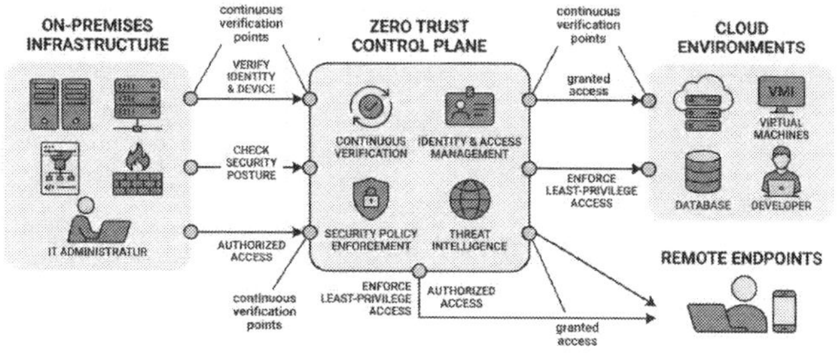

A unified control plane continuously verifies identity and security posture to grant least-privilege access across on-premises, cloud, and remote assets

9.5 Long-Term Program Design and Governance

Adaptive Strategic Planning: Objectives Without Rigid Blueprints

The adaptive planning architecture that the CISO in this chapter's opening scenario constructed has a specific structure that distinguishes it from traditional planning and from the absence of planning. At the highest level, it articulates strategic objectives that are stable enough to provide planning direction for two to three years: achieve zero-trust identity architecture across all critical systems, build organizational capability to evaluate and govern AI security tools, and establish an incident response capability that contains AI-augmented attacks within defined recovery time objectives. These objectives do not specify technologies, vendors, or implementation approaches. They define the destination without prescribing the route, which preserves the flexibility to adapt the route as conditions change while maintaining enough stability to guide multi-year investment.

Below the objectives, a rolling twelve-month investment plan specifies the initiatives, resources, and milestones that represent the current best path toward those objectives. This plan is reviewed quarterly, with explicit consideration of whether the assumptions that justified each initiative still hold. When the development of a new threat has invalidated an assumption, a vendor change, a technology shift, or the discovery of an organizational capability, the initiative is adjusted or replaced. The quarterly review is not a bureaucratic ritual. It is the mechanism by which the organization exercises strategic judgment in response to new information, maintaining coherence without turning the plan into a straitjacket that prevents rational responses to a changing environment. The discipline is not in writing the plan. It is in the willingness to change when the evidence demands it.

Metrics Architecture for Measuring Resilience, Not Just Compliance

The metrics that have traditionally dominated security program measurement, such as patch compliance rates, vulnerability counts, audit findings, and training completion percentages, measure activity, not resilience. They tell you whether the security team followed the checklist. They do not tell you whether the organization is actually better positioned to resist and recover from AI-augmented attacks than it was last quarter. A metrics architecture designed to measure resilience requires different measures: the time from initial compromise to detection, the time from detection to containment, the fraction of incidents that involved AI-augmented techniques, the fraction of red team scenarios that successfully exfiltrated data or achieved lateral movement, and the fraction of analyst overrides of AI recommendations that turned out to be correct. These metrics require investment in measurement infrastructure that most programs have not built, which is itself a meaningful readiness indicator.

Resilience metrics are harder to collect and harder to interpret than compliance metrics. They require instrumented incident response processes that generate consistent data across events. They require red team programs that produce comparable results across exercises. They require organizational maturity to report failure rates accurately without making those reports blame exercises that discourage honest measurement. Leaders who build the discipline to measure resilience rather than compliance will develop a more accurate picture of their organization's security posture and more defensible investment recommendations than leaders who rely on metrics that measure effort rather than outcome. The board presentation showing a trend line of decreasing mean time to contain incidents is more informative than one showing that patch compliance is at 94 percent.

The translation of resilience metrics into board-level reporting is an underappreciated skill. Boards are not equipped to interpret mean time to detection in technical terms or to evaluate the significance of a red team finding without context. They are equipped to understand trend direction, relative performance against industry benchmarks, and the relationship between investment and outcome. Security leaders who can translate operational resilience metrics into those terms demonstrating that the program has reduced the window of adversary advantage over the past year, that the organization's posture compares favorably to sector peers in specific domains, and that the investments the board has approved are producing measurable improvement in defined outcome metrics will sustain the executive support that AI-era security programs require over time.

Executive and Board Engagement Models for Sustained AI Security Investment

Sustained investment in AI security capabilities requires sustained executive conviction that the investment is warranted. A single compelling presentation does not maintain that conviction. It is maintained through a consistent, credible reporting relationship that provides executives with the information they need to make informed resource-allocation decisions quarter after quarter. The reporting relationship requires regularity, not only briefings at the moment of crisis or during the annual budget cycle, but also standing visibility into the security program's performance and direction. It requires calibrated communication that is accurate about limitations and failures, not just successes, because executives who receive only positive information cannot trust the assessment they are given. And it requires connection to business outcomes, not just technical metrics, but an explanation of what those metrics mean for the organization's ability to pursue its mission without disruption.

The specific cadence and format of executive engagement vary by organization. Some boards have dedicated technology or risk committees with the expertise and appetite for detailed security briefings. Others need the core message distilled into a small number of indicators they can interpret without a deep technical background. What is consistent across effective engagement models is that the security leader has invested in understanding what the executives and board need to know, in what form, and has designed a communication approach that meets that need rather than one that delivers the information that is easiest for the security team to produce. The investment in communication is as important as the investment in security capability, because capability without executive support cannot be sustained at the scale that AI-era threats require.

Diagram 9.5: Adaptive Security Program Architecture

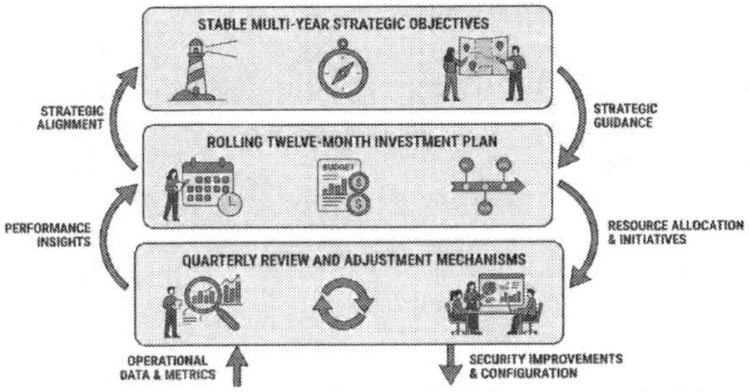

This adaptive architecture continuously optimizes security investments based on operational feedback and long-term strategic goals.

9.6 Manager's Future-Readiness Checklist

Replace the static roadmap with an adaptive planning architecture. Define two or three strategic objectives that express the desired security outcomes in terms that do not prescribe specific technologies or vendors. Establish a rolling twelve-month investment plan with a formal quarterly review process that includes explicit assessment of whether the assumptions underlying each initiative still hold. Document the assumptions so that they can be evaluated and challenged, not left implicit and invisible.

Build the feedback mechanisms that connect operational data to strategic decisions. Ensure that after-action reviews generate structured data that can be aggregated and analyzed for investment signals. Establish a formal process to translate incident pattern analysis into investment recommendations and deliver them to the appropriate decision-making forum. Do not allow operational learning to remain isolated in the teams that generated it, where it informs only local improvements rather than program-level investment.

Conduct a zero trust maturity assessment using CISA's Zero Trust Maturity Model or an equivalent framework. Identify the three highest-priority gaps relative to the AI-augmented attack techniques most relevant to the organization's risk profile. Build remediation of those gaps into the twelve-month investment plan with specific milestones, resource assignments, and accountability. Revisit the assessment annually and after any significant architectural change.

Review the threat intelligence program for strategic completeness. Assess whether intelligence consumption extends beyond tactical indicators to strategic insight that informs investment decisions. Identify whether the organization's participation in sector sharing programs is generating genuine operational value. Address gaps with specific resource and process investments rather than nominal membership adjustments.

Redesign executive and board security reporting to include resilience metrics alongside compliance metrics. Develop a communication framework that translates operational performance data into business-relevant terms. Establish a regular reporting cadence that maintains executive visibility between crisis moments. Invest in the communication discipline that sustained investment requires; it is not a soft skill ancillary to the technical work. It is the mechanism through which the technical work gets resourced year after year.

Diagram 9.6: Resilience Metrics Dashboard for Executive Reporting

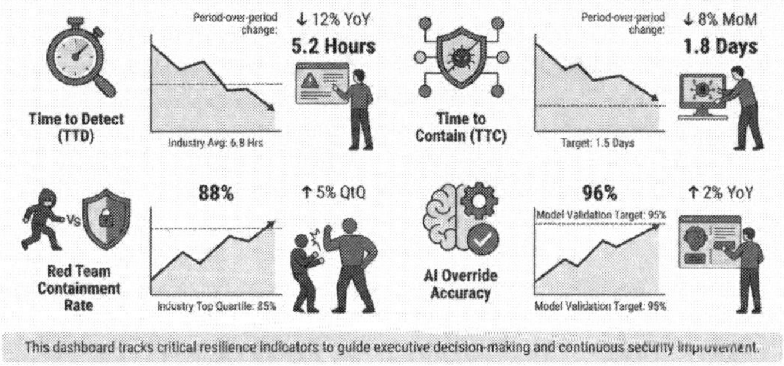

9.7 The Discipline of Adaptive Resilience

The CISO in this chapter's opening scenario made a decision that is counterintuitive in most organizational cultures: she replaced a plan with a practice. The plan was tangible, comprehensive, and politically reassuring. The practice was demanding, continuous, and required sustained leadership attention, without the comfort of a document declaring the strategy complete. The plan was obsolete before the organization had executed its first phase. The practice produced an organization that could respond to the AI-driven changes in the threat environment as they occurred rather than after they had compounded into a strategic deficit.

Architecture for an uncertain future is not primarily technological architecture. It is an organizational architecture: the planning practices, the feedback mechanisms, the intelligence integration, the metrics frameworks, and the governance structures that allow a security program to adapt continuously without losing strategic coherence. These structures do not eliminate uncertainty. They make the organization capable of functioning effectively in the presence of uncertainty, which is the most that any security program in the AI era can honestly promise.

Every chapter of this book has argued, in different contexts, for the same underlying discipline: deliberate human judgment applied to consequential decisions, sustained by rigorous process, governed by clear accountability, and continuously improved by honest measurement. That discipline is what separates security programs that advance their organizations' mission from programs that consume resources and provide false assurance. In the age of intelligent threats, the organizations that apply that discipline will be resilient. Those who do will not be protected by their technology alone.

10 The Leaders This Moment Requires

The organizations that navigate the AI cybersecurity era most successfully will not be those with the largest security budgets or the most sophisticated tools. They will be those led by executives who understand the nature of the conflict deeply enough to make the right bets, build the right teams, and sustain the organizational commitment that genuine resilience demands. That is a leadership challenge, not merely a technical one, and this book has been written in the conviction that the leaders who take it seriously can make a decisive difference.

What has changed most profoundly is the pace. The adversary is no longer static. Every defensive capability deployed today will be studied, probed, and eventually circumvented by AI systems that are themselves learning. This does not mean that defense is futile; it means that defense must be dynamic, and that the organizational structures supporting it must be built for adaptation rather than for a fixed posture. The leaders who internalize this shift will approach every investment decision, every hiring decision, and every governance decision with the question of whether it increases or decreases the organization's capacity to evolve.

The governance dimension deserves particular emphasis as a closing thought because it is the area where senior leaders can make the most distinctive contribution. Technical teams can implement detection systems and run playbooks. Only leadership can define the ethical boundaries within which AI-powered defense operates, establish the accountability structures that ensure those boundaries are respected, and create the cultural conditions in which security professionals feel empowered to raise concerns when AI systems behave in ways that conflict with organizational values. Governance, in this sense, is not a constraint on effective security; it is a prerequisite for it.

Every chapter of this book has touched on the human dimension: the analysts whose judgment AI is designed to augment rather than replace, the teams whose skills must continuously evolve, the executives whose strategic choices determine whether AI serves the organization's security objectives or merely accumulates as technical debt in a portfolio of tools that no one fully understands. Attending to the human dimension is not soft leadership; it is the most pragmatic possible response to a threat environment that is, itself, increasingly designed to exploit human limitations at scale.

The intelligent threat is real, present, and accelerating. The organizations that defend themselves effectively against it will be those that treat AI cybersecurity not as a specialized function operating in isolation, but as a strategic capability woven into every dimension of enterprise leadership. The frameworks in this book are a starting point. The work of applying, adapting, and building on them belongs to every leader who accepts responsibility for the enterprise's resilience.

11 Conclusion

There is a moment, in every era of technological change, when the world shifts quietly beneath our feet. Most people don't notice it at first. The dashboards still light up. The meetings still happen. The systems still hum. But somewhere in the background, something fundamental has changed. The rules that governed yesterday no longer apply, and the assumptions that once felt safe begin to crack.

For cybersecurity leaders, that moment has already arrived.

Throughout this book, we have walked through a landscape sometimes violently transformed by artificial intelligence. We have seen attackers who no longer wait, who no longer guess, who no longer rely on human patience or human error. We have seen malware that learns, phishing campaigns that speak in familiar voices, and synthetic media that can fracture trust with a single fabricated sentence. We have seen defenders racing to keep pace, building systems that analyze behavior, triage alerts, and respond at machine speed. And we have seen organizations grappling with the uncomfortable truth that the threat is no longer "out there." It is woven into the very technologies that power modern enterprise.

But beneath the technical detail, beneath the diagrams and frameworks, something deeper has been unfolding: **a story about leadership in a world where human and artificial intelligence collide.**

This is not a story about tools. It is a story about people.

It is the story of a SOC analyst staring at a screen full of alerts, knowing that somewhere in that noise is the one signal that matters. It is the story of a CISO walking into a boardroom, carrying the weight of a threat landscape that changes faster than budgets do. It is the story of a federal program manager trying to secure a system built twenty years ago against an adversary built in twenty minutes. And it is the story of every leader who has ever felt the quiet pressure of responsibility, the knowledge that the decisions they make today will determine whether their organization survives tomorrow.

AI has not made that responsibility easier. It has made it heavier.

But it has also made it more meaningful.

Because in a world where attackers move at machine speed, the organizations that endure will not be the ones with the biggest budgets or the most tools. They will be the ones with leaders who understand what this moment demands: clarity, courage, and a willingness to rethink everything they thought they knew about security.

- Leaders who understand that governance is not bureaucracy, it is the architecture of trust.
- Leaders who understand that talent is not a cost, it is the engine of resilience.

- Leaders who understand that adaptation is not a reaction, it is a discipline.
- Leaders who understand that AI is not a threat or a savior, it is a force multiplier, shaped by the hands that wield it.

Machines will not write the future of cybersecurity. It will be written by the people who learn how to lead alongside them.

As you close this book, imagine the enterprise you are responsible for. Imagine its systems, its people, its mission. Imagine the adversaries who are already probing its edges. And then imagine the version of that enterprise that is ready not just for the threats of today, but for the ones that have not yet been invented.

That future is not built in a single decision. It is built in a thousand small ones:

- A budget reallocated toward automation.
- A governance framework updated before it becomes necessary.
- A training program launched before the skill gap becomes a crisis.
- A culture shaped intentionally, not accidentally.
- A team empowered to think, question, and adapt.

This is the work of leadership in the age of intelligent threats. It is not glamorous. It is not simple. It is not quick.
But it is necessary. And it is possible.

The machines may move faster now. But leadership still determines the direction.

And in that truth lies the defender's greatest advantage.

The next chapter of cybersecurity will be written by those who choose to lead, not from fear or compliance, but from conviction. The conviction that resilience is achievable. That trust can be rebuilt. That teams can evolve. That organizations can adapt. And that intelligence, human and artificial, can be harnessed not just to defend the enterprise, but to strengthen it.

The age of intelligent threats has arrived.
So has the age of intelligent defense.
The question that remains is simple:

What kind of leader will you choose to be?

---

12 References

Allen, Julia, et al. "AI-Driven Malware: Trends in Adaptive Offensive Capabilities." *ACM Transactions on Privacy and Security*, vol. 27, no. 2, 2024. (CH1)

Baker, Thomas, and Elena Markov. "Reinforcement Learning in Modern Cyberattacks." *IEEE Security & Privacy*, vol. 21, no. 5, 2023. (CH1)

Bansal, Rishi, et al. "Adversarial Machine Learning in Enterprise Security." *Journal of Cybersecurity*, vol. 10, no. 1, 2024. (CH3)

Bergstrom, Peter, and Nadia Karim. "Synthetic Identity Fraud in the Age of AI." *Financial Crime Review*, vol. 18, no. 3, 2023. (CH4)

Brennan, Lisa. "Deepfake-Driven Social Engineering Attacks." *Harvard Business Review*, 2024. (CH4)

CISA, NSA, and FBI. *Contextualizing Deepfake Threats to Organizations.* U.S. Government, 2023. (CH4)

CISA. *Securing AI Systems: Best Practices for Defensive Deployment.* U.S. Cybersecurity and Infrastructure Security Agency, 2024. (CH5)

Cohen, I. Glenn, et al. "Regulating AI in Security-Critical Environments." *Health Affairs*, vol. 43, no. 2, 2024. (CH7)

Cybersecurity & Infrastructure Security Agency. *AI Security: Executive Framework for Risk Reduction.* CISA, 2025. (CH7)

European Union Agency for Cybersecurity (ENISA). *AI Threat Landscape 2023.* ENISA, 2023. (CH1)

European Union Agency for Cybersecurity (ENISA). *Securing Machine Learning Systems.* ENISA, 2024. (CH5)

Federal Trade Commission. *Combatting AI-Enabled Fraud and Deception.* FTC, 2024. (CH4)

GAO. *Science & Tech Spotlight: Combating Deepfakes.* U.S. Government Accountability Office, 2024. (CH4) U.S. Government Accountability Office (U.S. GAO)

Google DeepMind. "Adversarial Robustness in Large-Scale Models." *Nature Machine Intelligence*, vol. 6, 2024. (CH3)

IBM Security. *AI-Powered SOC Transformation Report, 2024, IBM.* (CH6)

Kaspersky Global Research. "Adaptive Malware and AI-Driven Ransomware Trends." *Kaspersky Security Bulletin*, 2023. (CH1)

Khan, Omar, et al. "Model Poisoning Attacks in Federated Learning." *IEEE Transactions on Neural Networks and Learning Systems*, 2023. (CH3)

Kumar, Sandeep, and Priya Natarajan. "AI-Enhanced SOC Workflows." *Journal of Information Security*, vol. 19, no. 4, 2024. (CH6)

McAfee Labs. *AI-Generated Phishing and Social Engineering Report 2024.* McAfee, 2024. (CH1)

Microsoft Security Response Center. *AI-Driven Threat Intelligence Review 2025.* Microsoft, 2025. (CH9)

MITRE. *Adversarial ML Threat Matrix 2.0.* MITRE Corporation, 2023. (CH3)

MITRE. *AI Security Architecture Principles.* MITRE, 2024. (CH5)

National Institute of Standards and Technology. *NIST AI Risk Management Framework (AI RMF 1.0).* NIST, 2023. (CH7)

National Institute of Standards and Technology. *Secure Software Development Framework for AI Systems.* NIST, 2024. (CH5)

National Security Agency. *Deepfake Threats to National Security.* NSA, 2023. (CH4) CISA National Security Agency

Norton Labs. *Synthetic Media and Identity Fraud Report 2024.* Norton, 2024. (CH4)

OpenAI. "Safety Challenges in Large-Scale AI Deployment." *OpenAI Research Papers*, 2023. (CH7)

Palo Alto Networks Unit 42. *AI-Enhanced Threat Landscape Report 2024.* Palo Alto Networks, 2024. (CH1)

PwC. *AI Governance and Cybersecurity: Executive Priorities for 2025.* PwC, 2025. (CH7)

SANS Institute. *AI-Driven SOC Modernization Playbook.* SANS, 2024. (CH6)

Symantec Threat Intelligence. "Self-Modifying Malware in AI-Augmented Campaigns." *Symantec Security Insights*, 2023. (CH1)

Trend Micro Research. *AI-Powered Cybercrime Ecosystems.* Trend Micro, 2024. (CH1)

World Economic Forum. *Global Cybersecurity Outlook 2024.* WEF, 2024. (CH9)

World Economic Forum. *AI Governance in Critical Infrastructure.* WEF, 2025. (CH7)

Made in the USA
Monee, IL
10 November 2026

b7c518de-bd49-4dbd-881e-2648db9b3373R01